COLLEGE PLACEMENT TEST
MATH PRACTICE
ADVANTAGE+ EDITION

350 College Placement Test
Math Practice Problems and Solutions

College Placement Test and CPT are registered trademarks. This publication is not endorsed by or affiliated with the official College Placement Test, nor with the College Board, the trademark owner of CPT.

College Placement Test Math Practice Advantage Plus Edition: 350 College Placement Test Math Practice Problems and Solutions

Copyright © 2007, 2011. Academic Success Media © COPYRIGHT 2020. Academic Success Group.

All rights reserved. No part of this publication may be reproduced, stored in a retrieval system, or transmitted, in any form or by any means, electronic, mechanical, photocopying, recording or otherwise.

ISBN: 978-1-949282-49-8

COPYRIGHT NOTICE TO EDUCATORS: Please respect copyright law! Under no circumstances may you make copies of these materials for distribution to or use by students. Should you wish to use the materials with students, you are required to purchase a copy of this publication for each of your students.

Note: College Placement Test and CPT are registered trademarks. This publication is not endorsed by or affiliated with the official College Placement Test, nor with the College Board, the trademark owner of CPT.

TABLE OF CONTENTS

COLLEGE PLACEMENT TEST MATH FORMULAS AT A GLANCE:

Absolute value	1
Combinations and permutations	1
Coordinate geometry	1
Distance formula	1
Exponent laws	2
Fractions containing fractions	2
Imaginary numbers	3
Logarithmic functions	3
Matrices and determinants	3
Quadratic formula	4
Geometric reasoning	4
Trigonometric reasoning	5

COLLEGE PLACEMENT TEST MATH PRACTICE	6
STEP-BY-STEP SOLUTIONS	21

ADVANTAGE PLUS EDITION – BONUS MATERIAL

150 Additional Math Problems with Tips and Formulas:

Arithmetic Problems	82
Geometry Problems	90
Quantitative Reasoning and Statistics	98
Algebra and Functions	103
Answers and Solutions for the Bonus Exercises	116
Math Formula Sheet	138

COLLEGE PLACEMENT TEST MATH FORMULAS AT A GLANCE:

Absolute value:

Absolute value is always a positive number. $|-x| = x$

Combinations and permutations:

Combinations:

To determine the number of combinations of S at a time that can be made from a set containing N items, you need this formula:

$(N!) \div [(N - S)! \times S!]$

Permutations:

In order to calculate the number of permutations of size S taken from N items, you should use this formula:

$N! \div (N - S)!$

Coordinate geometry:

Midpoints:

For two points on a graph (x_1, y_1) and (x_2, y_2), the midpoint is: $(x_1 + x_2) \div 2$, $(y_1 + y_2) \div 2$

Slope:

In order to calculate the slope of a line, you need this formula: $y = mx + b$

m is the slope and b is the y intercept (the point at which the line crosses the y axis).

If two lines are perpendicular, the product of their slopes is equal to –1. If two lines are parallel, they will have the same slope.

Distance formula:

$$d = \sqrt{(x_2 - x_1)^2 + (y_2 - y_1)^2}$$

Exponent laws:

Same base numbers with different exponents:

If the base number is the same, and the problem asks you to multiply, you simply add the exponents. If the base number is the same, and the problem asks you to *divide*, you *subtract* the exponents.

Example: $4^2 \times 4^5 = 4^{(2+5)} = 4^7$

Different base numbers:

If the base numbers are different, you need to multiply the base number, but add the exponents.
Example: $(2x^3)(-4x^5) = -8x^8$

Fractions as exponents:

Place the base number inside the radical sign. The denominator of the exponent is the nth root of the radical, and the numerator is new exponent. Example: $x^{2/5} = (\sqrt[5]{x})^2$

Negative exponents:

Remove the negative sign on the exponent by expressing the number as a fraction, with 1 as the numerator. Example: $x^{-3} = \dfrac{1}{x^3}$

Fractions containing fractions:

When you see fractions that have fractions within themselves, remember to treat the denominator as the division sign.

Example: $\dfrac{x + \dfrac{1}{x}}{\dfrac{1}{x}} = ?$

$\dfrac{x + \dfrac{1}{x}}{\dfrac{1}{x}} =$

$\left(x + \dfrac{1}{x}\right) \div \dfrac{1}{x} =$

Then invert and multiply the fractions as usual. In this case $\frac{1}{x}$ becomes $\frac{x}{1}$ when inverted, which is then simplified to x.

$$\left(x+\frac{1}{x}\right) \div \frac{1}{x} =$$

$$\left(x+\frac{1}{x}\right) \times x = x^2 + \frac{x}{x} = x^2 + 1$$

Imaginary numbers:

Remember that two complex numbers are equal only if their real parts equal one another and their imaginary parts also equal one another.

Example: *a* and *c* are real numbers. *xi* and *yi* are imaginary numbers. When does a + xi = c + yi ?

Since two complex numbers are equal if and only if their real parts are equal and their imaginary parts are equal, *a* must be equal to *c* and *xi* must be equal to *yi*.

Logarithmic functions:

Logarithmic functions are just another way of expressing exponents.

$x = \log_y Z$ is always the same as: $y^x = Z$

Matrices and determinants:

In order to find the determinant for a two-by-two matrix, you need to cross multiply and then subtract:

Example: What is the determinant of the following matrix?

$$\begin{bmatrix} a & b \\ c & d \end{bmatrix}$$

To find the determinant, first *a* is multiplied by *d* and *c* is multiplied by *b*.

Then we subtract the two terms to get the determinant: ad – cb = ad – bc

Quadratic formula:

For any equation in this form: $ax^2 \pm bx \pm c = 0$, the quadratic formula is as follows:

$x = (-b \pm \sqrt{b^2 - 4ac})/2a$

Trigonometric and geometric reasoning:

Geometric reasoning questions cover the measurement of circles, triangles, cylinders, cones, or parallelograms.

<u>Area of circles:</u> π × R^2 (radius squared)

<u>Area of squares and rectangles:</u> length × width

<u>Area of triangles:</u> (base × height) ÷ 2

<u>Arcs:</u> Arc length is the distance on the outside (or circumference) of a circle.

<u>Chords:</u> Chord length is always the straight line connecting the given points.

<u>Circumference of a circle:</u> π × diameter (diameter = radius × 2)

<u>Cone volume:</u> (π × radius² × height) ÷ 3

<u>Cylinder volume:</u> $V = \pi r^2 h$

In other words, to calculate the volume of a cylinder you take π times the radius squared times the height.

<u>Hypotenuse length:</u> hypotenuse length $C = \sqrt{A^2 + B^2}$

In other words, the length of the hypotenuse is always the square root of the sum of the squares of the other two sides of the triangle:

<u>Perimeter of squares and rectangles:</u> (length × 2) + (width × 2)

<u>Radians:</u> θ = s ÷ r

θ = the radians of the subtended angle
s = arc length
r = radius

π ÷ 6 × radians = 30°
π ÷ 4 × radians = 45°
π ÷ 2 × radians = 90°

π × radians = 180°
π × 2 × radians = 360°

Tangent, sine, and cosine:

The following trigonometric rules are valid with respect to any angle:

$\sin A^2 = 1 - \cos A^2$
$\cos A^2 = 1 - \sin A^2$
$\cos A^2 + \sin A^2 = 1$

Here are important trigonometric formulas for calculating the sine, cosine, and tangent of any given angle A, as in the illustration:

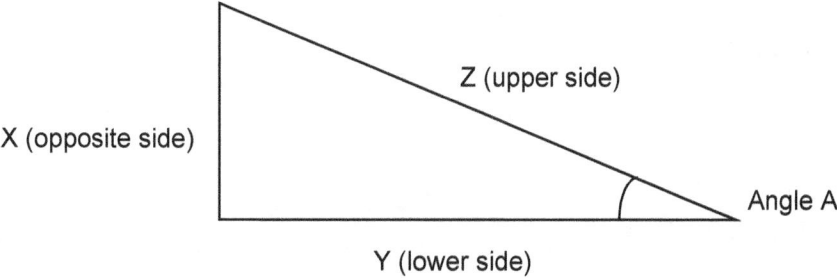

$\sin A = x/z$
$\cos A = y/z$
$\tan A = x/y$

In other words, for any angle, sine is calculated by taking the measurement of the opposite side divided by the measurement of the upper side of the triangle.

Cosine is calculated by taking the measurement of the lower side divided by the measurement of the upper side of the triangle.

Tangent is calculated by taking the measurement of the opposite side divided by the measurement of the lower side of the triangle.

COLLEGE PLACEMENT TEST MATH PRACTICE

1) 7.23 + .004 + .513 = ?

2) 8.13 × 3.1 = ?

3) Two people are going to give money to a foundation for a project. Person A will provide one-half of the money. Person B will donate one-eighth of the money. What fraction represents the unfunded portion of the project?

4) Which of the following is the least?
A) .32
B) .032
C) .23
D) .302

5) Convert the following to decimal format: $3/20$

6) 60 is 20 percent of what number?

7) $6¾ - 2½$ = ?

8) Estimate the result of the following: 12.9 × 3.1

9) Express 30 percent of y as a fraction with 100 as the denominator.

10) 152 ÷ 8 = ?

11) 7.55 + .055 + .02 = ?

12) Estimate the result of the following: 502 ÷ 4.9

13) Beth took a test that had 60 questions. She got 10% of her answers wrong. How many questions did she answer correctly?

14) Which of the following is the greatest?
A) .540
B) .054
C) .045
D) .5045

15) 4.602 – 0.32 = ?

16) 4.27 × 3.1 = ?

17) A job is shared by 4 workers, A, B, C, and D. Worker A does 1/6 of the total hours. Worker B does 1/3 of the total hours. Worker C does 1/6 of the total hours. What fraction represents the remaining hours allocated to person D?

18) 120 students took a math test. The 60 female students in the class had an average score of 95, while the 60 male students in the class had an average of 90. What is the average test score for all 120 students in the class?

19) $3\frac{1}{2} - 2\frac{2}{5} = ?$

20) .18 ÷ .06 = ?

21) $\frac{1}{32}$ is equivalent to what percentage?

22) Yesterday the temperature was 90 degrees. Today it is 10% cooler than yesterday. What is today's temperature?

23) 1/3 – 1/7 = ?

24) A class contains 20 students. On Tuesday 5% of the students were absent. On Wednesday 20% of the students were absent. How many more students were absent on Wednesday than on Tuesday?

25) Farmer Brown owns a herd of cattle. This year, his herd consisted of 250 cows. Then he sold 60% of his herd. How many cows did he sell?

26) Estimate the result of the following: $30\frac{1}{4} \times 8\frac{9}{10}$

27) 6.55 × 1.1 = ?

28) Three people are going to contribute money to a charity. Person A will provide one-third of the money. Person B will contribute one-half of the money. What fraction represents Person C's contribution of money for the project?

29) The snowfall for November is 5 inches more than for December. If the total snowfall for November and December is 35 inches, what was the snowfall for November?

30) 2/3 – 1/6 = ?

31) A museum counts its visitors each day and rounds each daily figure up or down to the nearest 10 people. 104 people visit the museum on Monday, 86 people visit the museum on Tuesday, and 81 people visit the museum on Wednesday. What amount best represents the number of visitors to the museum for the three days, after rounding?

32) Convert the following fraction into decimal format: $\frac{4}{50}$

33) Mount Pleasant is 15,138 feet high. Mount Glacier is 9,927 feet high. What is the best estimate of the difference between the altitudes of the two mountains to the nearest thousand?

34) John is measuring plant growth as part of a botany experiment. Last week, his plant grew 7¾ inches, but this week his plant grew 10½ inches. By how much did this week's growth surpass last week's?

35) At the beginning of a class, one-fourth of the students leave to attend band practice. Later, one half of the remaining students leave to go to PE. If there were 15 students remaining in the class at the end, how many students were in the class at the beginning?

36) $5\frac{1}{3} - 1\frac{1}{4} = ?$

37) $2/10$ is equivalent to what percentage?

38) Estimate the following: $201 \div 3.9$

39) $1.25 + .655 + .002 = ?$

40) Tom bought a shirt on sale for $12. The original price of the shirt was $15. What was the percentage of the discount on the sale?

41) Shania is entering a talent competition which has three events. The third event (C) counts three times as much as the second event (B), and the second event counts twice as much as the first event (A). What equation, expressed only in terms of variable A, can be used to calculate Shania's final score for the competition?

42) $6 \div 40 = ?$

43) $(-12 + 6) \div 3 = ?$

44) $(x^2 - 4) \div (x + 2) = ?$

45) If $5x - 2(x + 3) = 0$, then $x = ?$

46) Simplify the following equation: $(x + 3y)^2$

47) $(x + 3y)(x - y) = ?$

48) What is the value of the expression $6x^2 - xy + y^2$ when $x = 5$ and $y = -1$?

49) Two people are going to work on a job. The first person will be paid $7.25 per hour. The second person will be paid $10.50 per hour. If A represents the number of hours the first person will work, and B represents the number of hours the second person will work, what equation represents the total cost of the wages for this job?

50) If a circle A has a radius of 4, what is the circumference of the circle?

51) $8^7 \times 8^3 = ?$

52) $20 - \frac{3}{4}X > 17$, then $X < ?$

53) $-6(4 - 1) - 2(5 - 2) = ?$

54) Simplify: $|3 - 6|$

55) Express the following number in scientific notation: 625

56) $\sqrt{5}$ is equivalent to what number in exponential notation?

57) State the x and y intercepts that fall on the straight line represented by the following equation:

$y = x + 6$

58) $(5x + 7y) + (3x - 9y) = ?$

59) Simplify the following: $(5x^2 + 3x - 4) - (6x^2 - 5x + 8)$

60) $(x - 4)(3x + 2) = ?$

61) Simplify: $\sqrt{7} + 2\sqrt{7}$

62) Factor the following: $x^2 + x - 20$

63) $(-5 - (-14)) \div 2 = ?$

64) Mark's final grade for a course is based on the grades from two tests, A and B. Test A counts toward 35% of his final grade. Test B counts toward 65% of his final grade. What equation is used to calculate Mark's final grade for this course?

65) $(x - 4y)^2 = ?$

66) If $4x - 3(x + 2) = -3$, then $x = ?$

67) $(x^2 - x - 12) \div (x - 4) = ?$

68) $(3 + -13) \div 5 = ?$

69) If A represents the number of apples purchased at 20 cents each and B represents the number of bananas purchased at 25 cents each, what equation represents the total value of the purchase?

70) If circle A has a radius of 0.4 and circle B has a radius of 0.2, what is the difference in area between the two circles?

71) What is the value of the expression $2x^2 + 3xy - y^2$ when $x = 3$ and $y = -3$?

72) $(x - y)(3x + y) = ?$

73) $\sqrt{2} \times \sqrt{3} = ?$

74) $-3(5 - 2) - 6(4 - 3) = ?$

75) $20 - \frac{4}{5}X > 16$, then $X < ?$

76) $(-12 + 8) \div 2 = ?$

77) If $7x - 5(x + 1) = -3$, then $x = ?$

78) Simplify: $(x - 2y)(2x - y)$

79) $(2x - y)(x - 3y) = ?$

80) What is the value of the expression $3x^2 - xy + y^2$ when $x = 2$ and $y = -2$?

81) If a circle has a radius of 6, what is the circumference of the circle?

82) $20 - \frac{1}{4}X > 18$, then $X < ?$

83) $-5(3 - 1) - 2(5 - 7) = ?$

84) $3^4 \times 3^3 = ?$

85) $(2x + 5y)^2 = ?$

86) $-2(4 - 1) - 4(3 - 2) = ?$

87) If $5x - 4(x + 2) = -2$, then $x = ?$

88) $(-10 + 1) \div 3 = ?$

89) What is the value of the expression $x^2 - xy + y^2$ when $x = 4$ and $y = -3$?

90) If a circle has a diameter of 18, what is the circumference of the circle?

91) If $x - 2(x + 3) = -8$, then $x = ?$

92) Simplify: $(x - y)(x + y)$

93) $\sqrt{8} \times \sqrt{2} = ?$

94) Factor the following: $2xy - 8x^2y + 6y^2x^2$

95) $(x + 3) - (4 - x) = ?$

96) What number is next in this sequence? 2, 4, 8, 16

97) $5^8 \div 5^2 = ?$

98) Simplify the following: $(4x^2 - 5x - 3) - (x^2 + 10x)$

99) A car travels at 60 miles per hour. The car is currently 240 miles from Denver. How long will it take for the car to get to Denver?

100) How many 3 letter permutations can be made from the following five letter set: F U N K Y?

101) If $x - 1 > 0$ and $y = x - 1$, then $y > $?

102) What is the determinant of the following matrix: $\begin{bmatrix} j & k \\ m & n \end{bmatrix}$

103) Find the coordinates (x, y) of the midpoint of the line segment on a graph that connects the points $(-5, 3)$ and $(3, -5)$.

104) The price of socks is $2 per pair and the price of shoes is $25 per pair. Anna went shopping for socks and shoes, and she paid $85 in total. In this purchase, she bought 3 pairs of shoes. How many pairs of socks did she buy?

105) Consider a two-dimensional linear graph where $x = 3$ and $y = 14$. The line crosses the y axis at 5. What is the slope of this line?

106) If $5 + 5(3\sqrt{x} + 4) = 55$, then $\sqrt{x} = $?

107) What ordered pair is a solution to the following system of equations?
$x + y = 11$
$xy = 24$

108) xi and yi are imaginary numbers. a and b are real numbers.

When does $xi - a = yi - b$?

109) Consider a right-angled triangle, where side M and side N form the right angle, and side L is the hypotenuse. If $M = 3$ and $N = 2$, what is the length of side L?

110) Express the equation $2^5 = 32$ as a logarithmic function.

111) $x^{-7} = $?

112) For the following equation, i represents an imaginary number. Simplify the equation:

$(2 - 2i) - (4 - 3i)$

113) $\dfrac{5x^3}{4} \times \dfrac{7}{x^2} = $?

114) $(-3x)(-6x^4) = $?

115) $\dfrac{x^2+8x+12}{x^2+8x+16} \times \dfrac{x^2+4x}{x^2+11x+30} = ?$

116) $\dfrac{3}{x^2+2x+1} + \dfrac{5}{x^2+x} = ?$

117) $3 = -\dfrac{1}{8}x$, then x = ?

118) $\sqrt{4x-4} = 6$, then x = ?

119) Write the slope-intercept equation for the following coordinates: (3,0) and (8,2)

120) 50 ÷ 5 + 36 ÷ 6 = ?

121) Calculate the slope and the y intercept: $3x + 5y = 24$

122) $\dfrac{7x+7}{x} \div \dfrac{4x+4}{x^2} = ?$

123) What figure should be placed inside the parentheses? $49x^8 = 7x(\ \ )$

124) $(4x^8 + 5x^5 - 7) - (-6x^5 + 5x^8 - 7) = ?$

125) $\dfrac{2}{3x} = \dfrac{?}{9x^2}$

126) Find the lowest common denominator and express as one fraction: $\dfrac{8}{x} + \dfrac{3}{x+2}$

127) $B = \dfrac{1}{3}CD$ Express in the following form: D =

128) What are two possible values of x for the following equation? $6x^2 + 16x + 8 = 0$

129) Simplify: $\dfrac{x^2}{x^{-8}}$

130) If $W = \dfrac{XY}{Z}$, then Z = ?

131) Find the volume of a cylinder whose height is 18 and whose radius is 4. Use 3.14 for π.

132) $\dfrac{2}{15x} - \dfrac{4}{21x^2} = ?$

133) $A = \dfrac{1}{2}(B+C)d$, if A = 120, B = 13, d = 8, then C = ?

134) Perform the operation: $10ab^5(5ab^7 - 4b^3 - 10a)$

135) The sum of twice a number and 8 less than the number is the same as the difference between −28 and the number. What is the number?

136) Perform the operation. Then simplify: $\dfrac{z^2 + 7z + 10}{z^2 + 13z + 40} \div \dfrac{z+8}{z^2 + 16z + 64}$

137) Simplify: $\dfrac{x + \dfrac{1}{x}}{\dfrac{1}{x}}$

138) If $\dfrac{3a}{10} + 9 = 12$, a = ?

139) Prepare the slope-intercept formula, using the data from the following table:

x	0	4	8
y	5	1	−3

140) If $\dfrac{20}{\sqrt{x^2 + 7}} = 5$, then x² = ?

141) Express as a rational number: $\sqrt[3]{\dfrac{64}{125}}$

142) In the standard (x,y) plane, what is the distance between (3,0) and (6,4)?

143) Give the slope-intercept formula that defines a line which is perpendicular to the line given by the formula: $y = \dfrac{1}{2}x + 5$

144) What is the value of a when $\dfrac{b^2 - ab + 24}{b - 12} = b - 2$?

145) $125^{-2/3} = ?$

146) If $\dfrac{18}{\sqrt{x^2 + 4}} = 6$, then $x = ?$

147) What equation defines a line that is parallel to the line given by the following equation: $y = -0.5x + 5$?

148) Rationalize the denominator: $\sqrt{\dfrac{16}{3}}$

149) What is the product of $(\sqrt{2} - 5\sqrt{5})$ and $(3\sqrt{2} - 4\sqrt{5})$?

150) For all $x \neq 0$ and $y \neq 0$, $\dfrac{4x}{1/xy} = ?$

151) $64^{3/2} = ?$

152) Simplify: $\dfrac{\sqrt{75}}{3} + \dfrac{5\sqrt{5}}{6}$

153) $\dfrac{\sqrt{36}}{3} + 5\dfrac{\sqrt{5}}{9} = ?$

154) $\sqrt{18} + 4\sqrt{75} + 5\sqrt{27} = ?$

155) The Smith family is having lunch in a diner. They buy hot dogs and hamburgers to eat. The hot dogs cost $2.50 each, and the hamburgers cost $4 each. They buy 3 hamburgers. They also buy hot dogs. The total value of their purchase is $22. How many hot dogs did they buy?

156) $(4x^2 + 3x + 5)(6x^2 - 8) = ?$

157) $13^3 \times 13^5 = ?$

158) Factor the following equation: $6xy - 12x^2y - 24y^2x^2$

159) If $x - 5 < 0$ and $y < x + 10$, then $y < ?$

160) $\sqrt{4x^8} \sqrt{6x^4} = ?$

161) What number is next in the sequence? 7, 14, 21, 28

162) Find the x and y intercepts of the following equation: $4x^2 + 9y^2 = 36$

163) Find the midpoint between the following coordinates: (2, 2) and (4, –6)

164) $-|10 - 17| = ?$

165) $\sqrt{-9} = ?$

166) Find the determinant of the following two–by–two matrix: $\begin{bmatrix} 4 & -1 \\ 3 & -2 \end{bmatrix}$

167) Convert $3^5 = 243$ to the equivalent logarithmic expression.

168) How many 2 letter combinations can be made from the five letter set: A B C D E?

169) Which one of the following is a solution to the following ordered pairs of equations:
$y = -2x - 1$
$y = x - 4$

A) (0, 1)
B) (1, 3)
C) (4, 0)
D) (1, –3)

170) Find the value of the following:

$$\sum_{x=2}^{4} x + 1$$

171) In the standard (x,y) plane, what is the distance between $(3\sqrt{5}, 0)$ and $(6\sqrt{5}, 4)$?

172) $\dfrac{a^3/ab}{b/5b^2} = ?$

173) A magician has a bag of colored scarves for a magic trick that he performs. The bag contains 3 blue scarves, 1 red scarf, 5 green scarves, and 2 orange scarves. If the magician removes scarves at random and the first scarf he removes is red, what is the probability that the next scarf will be orange?

174) **Use the chart below to answer the question that follows.**

X	Y
2	4
4	16
6	
8	64
10	100

The chart above shows the mathematical relationship between X and Y. What is the value of Y that is missing from the chart?

175) $(x^2 \div y^3)^3 = ?$

176) Find the area of the right triangle whose base is 2 and height is 5.

177) Consider the laws of sines and cosines for any given angle A. $\cos A^2 + \sin A^2 = ?$

178) Find the volume of a cone which has a radius of 3 and a height of 4.

179) $2^4 \times 2^2 = ?$

180) The perimeter of a rectangle is 64 meters. If the width were increased by 2 meters and the length were increased by 3 meters, what is the perimeter of the new rectangle?

181) $(-3x^2 + 7x + 2)(x^2 - 5) = ?$

182) $(A^5 \div A^2)^4 = ?$

183) If Д is a special operation defined by (x Д y) = (2x ÷ 4y) and (8 Д y) = 16, then y = ?

184) How many 3 letter permutations can be made from the four letter set: Z E B A?

185) Consider a right–angled triangle, where side A and side B form the right angle, and side C is the hypotenuse. If A = 5 and B = 3, what is the length of side C?

186) Consider the vertex of an angle at the center of a circle. If the diameter of the circle is 2, and if the angle measures 90 degrees, what is the arc length relating to the angle?

187) Pat wants to put wooden trim around the floor of her family room. Each piece of wood is 1 foot in length. The room is rectangular and is 12 feet wide and 10 feet long. How many pieces of wood does Pat need for the entire perimeter of the room?

188) The Johnson's have decided to remodel their upstairs. They currently have 4 rooms upstairs that measure 10 feet by 10 feet each. When they remodel, they will make one large room that will be 20 feet by 10 feet and two small rooms that will each be 10 feet by 8 feet. The remaining space is to be allocated to a new bathroom. What are the dimensions of the new bathroom?

189) In the figure below, x and y are parallel lines, and line z is a transversal crossing both x and y. Which three angles are equal in measure? (There are two possible answers.)

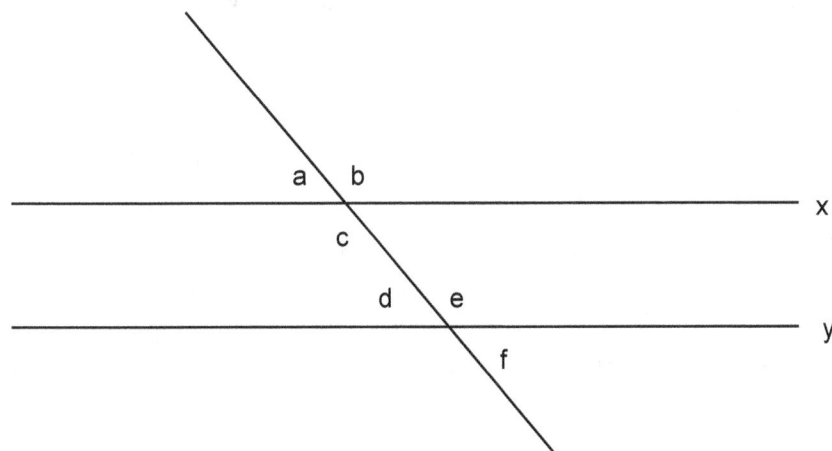

190) The central angle in the circle below measures 45° and is subtended by an arc which is 4π centimeters in length. How many centimeters long is the radius of this circle?

Arc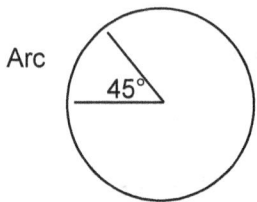

191) In the figure below, XY and WZ are parallel, and lengths are provides in units. What is the area of trapezoid WXYZ in square units?

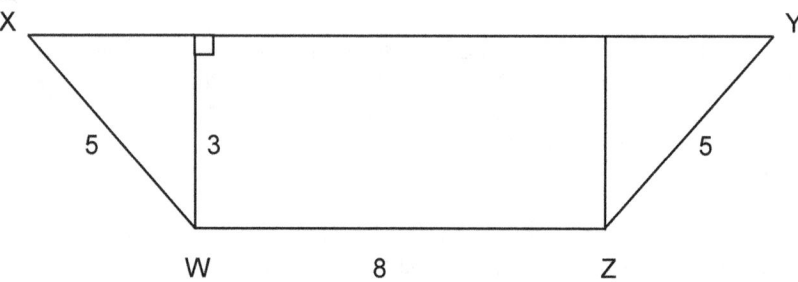

192) In the figure below, the lengths of KL, LM, and KN are provided in units. What is the area of triangle NLM in square units?

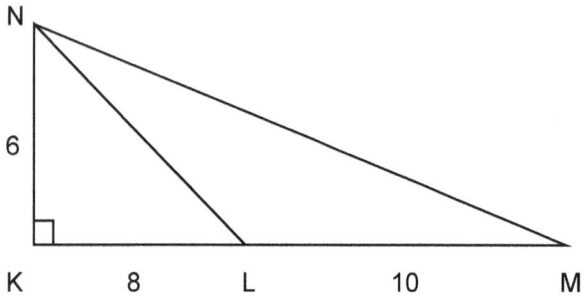

193) ∠XYZ is an isosceles triangle, where XY is equal to YZ. Angle Y is 30° and points W, X, and Z are co–linear. What is the measurement of ∠WXY?

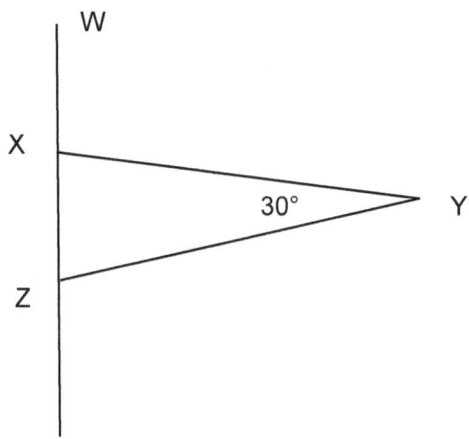

194) Consider the laws of sines and cosines with respect to angle A. $1 - \cos A^2 = ?$

195) If $\cos A = y/z$ and $\sin A = x/z$, then $\tan A = ?$

196) In the right triangle below, the length of AC is 5 units and the length of BC is 4 units. What is the tangent of ∠A ?

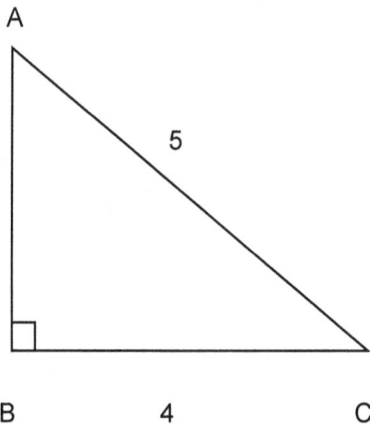

197) In the right angle in the figure below, the length of XZ is 10 units, sin 40° = 0.643, cos 40° = 0.776, and tan 40° = 0.839. Approximately how many units long is XY?

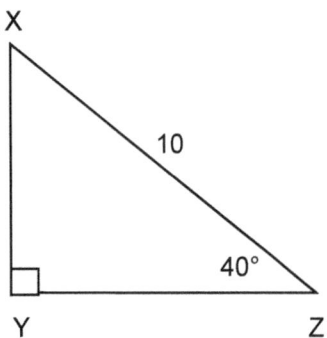

198) An arc length of θ on a circle of radius one subtends an angle of how many radians at the center of the circle?

199) If the radius of a circle is 1, what equation can be used to find the radians in 90°?

200) The street that runs between the hospital (H) and the police station (P) in the illustration below forms a 65° angle. If the police station (P) is 2.5 miles from the fire station (F), what trigonometric equation can be used to calculate the distance of the fire station from the hospital?

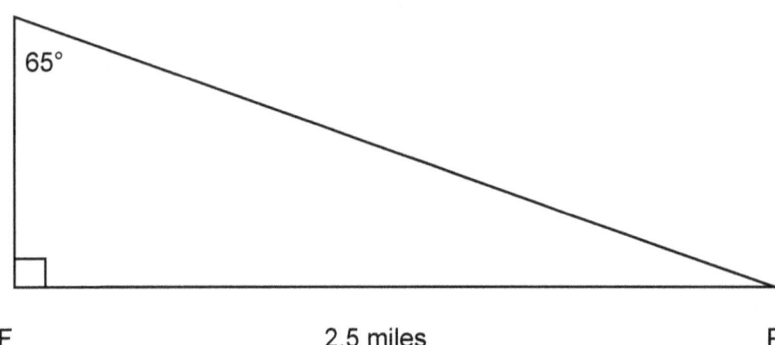

COLLEGE PLACEMENT TEST MATH PRACTICE – Solutions and Explanations:

1) The correct answer is: 7.747

When you add on your scratch paper, be sure to line all of the decimals up in a column like this:

```
7.230
0.004
0.513
─────
7.747
```

As you can see, you should add zeros where necessary at the beginning or end of the numbers in order to make the decimal points line up.

2) The correct answer is: 25.203

Tip: Be sure to put the decimal point in the correct position after you do the long multiplication.

We know that the decimal point has to be three places from the right on the final product because 8.13 has 2 decimal places and 3.1 has 1 decimal place, so 1 plus 2 equals 3 places.

```
   8.13
×   3.1
   ─────
    .813
  24.390
  ──────
  25.203
```

3) The correct answer is: 3/8

The sum of all contributions must be equal to 100%, simplified to 1. Let's say that the variable U represents the unfunded portion of the project. So the equation that represents this problem is: A + B + U = 1. Substitute with the fractions that have been provided: 1/2 + 1/8 + U = 1

Then find the lowest common denominator of the fractions.

4/8 + 1/8 + U = 1
5/8 + U = 1
U = 1 − 5/8
U = 3/8

4) The correct answer is: 0.032

Line all of the decimal points up for problems like this. Put in zeros where necessary, as follows:

0.320
0.032
0.230
0.302

When you have them lined up like this, you can see that 0.032 is the smallest one.

5) The correct answer is: 0.15

In order to convert a fraction to a decimal, you must do long division until you have no remainder.

```
      .15
20)3.00
    2.0
    1.00
    1.00
       0
```

6) The correct answer is: 300

20 percent is equal to 0.20. The phrase "of what number" indicates that we need to divide the two amounts given in the problem: 60 ÷ 0.20 = 300

We can check this result as follows: 30 × 0.20 = 60

7) The correct answer is: 4¼

Questions like this test your knowledge of mixed numbers. Mixed numbers are those that contain a whole number and a fraction. If the fraction on the first mixed number is greater than the fraction on the second mixed number, you can subtract the whole numbers and the fractions separately. Remember to use the lowest common denominator on the fractions.

6 – 2 = 4

3/4 – 1/2 = 3/4 – 2/4 = 1/4

Therefore, the result is 4¼.

8) The correct answer is: 39

For estimation problems like this, round the decimals up or down to the nearest whole number. 12.9 is rounded up to 13, and 3.1 is rounded down to 3. Then do long multiplication.

```
   13
×   3
   39
```

9) The correct answer is: 30y/100

This question tests your knowledge of how to express percentages as fractions.

Percentages can always be expressed as that number over one hundred. So 30% = 30/100.

10) The correct answer is: 19

You must do long division until you have no remainder.

```
    19
8) 152
    8
   ‾‾
   72
   72
   ‾‾
    0
```

11) The correct answer is: 7.625

Remember to line up the decimal points when you add.

7.550
0.055
0.020
‾‾‾‾‾
7.625

12) The correct answer is: 100

Remember to round the numbers up or down to the nearest whole number.

502 is rounded down to 500, and 4.9 is rounded up to 5.

Then divide: 500 ÷ 5 = 100

13) The correct answer is: 54

You must first determine the percentage of questions that Beth answered correctly.

We know that she got 10% of the answers wrong, so therefore the remaining 90% were correct.

Now we multiply the total number of questions by the percentage of correct answers.

60 × 90% = 54

14) The correct answer is: .540

Remember to put in zeros and line up the decimal points when you compare the numbers.

0.5400
0.0540
0.0450
0.5045

Therefore, the largest number is .540

15) The correct answer: 4.282

Subtract the numbers on your scratch paper, being sure to line the decimals up in a column.

```
  4.602
 −0.320
  4.282
```

16) The correct answer is: 13.237

Be careful with the decimal point positions when you do long multiplication.

```
    4.27
 ×   3.1
    .427
  12.810
  13.237
```

17) The correct answer is: 1/3

The sum of the work from all four people must be equal to 100%, simplified to 1. In other words, they make up the total hours by working together.

A + B + C + D = 1

1/6 + 1/3 + 1/6 + D = 1

Now, find the lowest common denominator of the fractions.

3 × 2 is 6. So the lowest common denominator is 6.

Now convert the fractions as required.

1/3 × 2/2 = 2/6

Now add the fractions together.

1/6 + 2/6 + 1/6 + D = 1
4/6 + D = 1
4/6 − 4/6 + D = 1 − 4/6
D = 1 − 4/6
D = 2/6 = 1/3

18) The correct answer is: 92.5

You need to find the total points for all the females and the total points for all the males. Then add these two amounts together and divide by the total number of students in the class to get your solution.

Females: 60 × 95 = 5700

Males: 60 × 90 = 5400

(5700 + 5400) ÷ 120 = 11,100 ÷ 120 = 92.5

19) The correct answer is: $1\frac{1}{10}$

Remember that if the fraction on the first mixed number is greater than the fraction on the second mixed number, you can subtract the whole numbers and the fractions separately.

3 – 3 = 1

1/2 – 2/5 = 5/10 – 4/10 = 1/10

Therefore, the result is $1\frac{1}{10}$

20) The correct answer is: 3

You must do long division until you have no remainder. Remember to line up the decimal points.

```
        3.0
.06) .18
     .18
       0
```

21) The correct answer is: 3.125%

1 ÷ 32 = 0.03125

0.03125 = 3.125%

22) The correct answer is: 81 degrees

If it is 10% cooler today, today's temperature is 90% of yesterday's temperature. So today's temperature is 90 degrees × 90% = 81 degrees

23) The correct answer is: 4/21

First, find the lowest common denominator.

1/3 × 7/7 = 7/21
1/7 × 3/3 = 3/21

When you have got both fractions in the same denominator, you subtract them.

7/21 – 3/21 = 4/21

24) The correct answer is: 3

Figure out the amount of absences for the two days and then subtract.

Tuesday's absences: 20 × 5% = 1

Wednesday's absences: 20 × 20% = 4

4 − 1 = 3

25) The correct answer is: 150

60% = .60

250 × .60 = 150

26) The correct answer is: 270

Remember to round the fractions up or down to the nearest whole number. Then do the multiplication.

30 × 9 = 270

27) The correct answer is: 7.205

Tip: Be sure to put the decimal point in the correct position after you do the multiplication. You can avoid long multiplication by removing and replacing the decimal points.

Remove the decimal points:

655 × 11 = (655 × 10) + (655 × 1) =
6550 + 655 = 7205

6.55 has a decimal point two places from the right. 1.1 has a decimal point 1 place from the right. So we know that we have to put the decimal point *three* numbers from the right on the final product of 7205. Therefore the final answer is 7.205

28) The correct answer is: 1/6

The three people make up the whole contribution by paying in together, so the sum of contributions from all three people must be equal to 100%, simplified to 1.

A + B + C = 1
1/3 + 1/2 + C = 1

Now, find the lowest common denominator of the fractions.

2/6 + 3/6 + C = 1
Therefore, C = 1/6

29) The correct answer is: 20 inches

Subtract the difference in snowfall between the two months from the total snowfall for the two months, and then divide by 2 in order to get the December snowfall.

35 − 5 = 30

30 ÷ 2 = 15

Now add back the excess for November to get the total for November.

15 + 5 = 20

30) The correct answer is: 1/2

First, find the lowest common denominator.

2/3 × 2/2 = 4/6

When you have got both fractions in the same denominator, you subtract them.

4/6 − 1/6 = 3/6, simplified to 1/2

31) The correct answer is: 270

Tip: A basic guideline for rounding is that 5 or more is rounded up, while 4 or less is rounded down.

Do the rounding for each day separately (before doing the addition) because this is stipulated in the problem. Then add together to solve the problem.

104 Rounded to 100

86 Rounded to 90

81 Rounded to 80

100 + 90 + 80 = 270

32) The correct answer is: 0.08

Do long division until you have no remainder.

```
       .08
   50)4.00
      4.00
         0
```

Alternatively, you know that 50 goes into 100 two times. So you can avoid long division by multiplying the numerator by 2 and adding a decimal point: 4 × 2 = 8% = .08

33) The correct answer is: 5,000 feet

Subtract the two amounts and then do the rounding. 15,138 − 9,927 = 5,211 (Rounded to 5,000)

Check by rounding the individual amounts as follows: 15,000 − 10,000 = 5,000

34) The correct answer is: 2¾ inches

This is essentially a mixed number problem. Here you can covert the fraction to decimals to make the subtraction easier.

10½ − 7¾ = 10.5 − 7.75 = 2.75 = 2¾

35) The correct answer is: 40 students

You need to create an equation to set out the facts of this problem. Here we will say that the total number of students is variable S.

15 = (S − ¼S) × ½

15 = ¾S × ½

15 = ³⁄₈S

15 × 8 = ³⁄₈S × 8

120 = 3S

S = 40

36) The correct answer is: 4¹⁄₁₂

If the fraction on the first mixed number is greater than the fraction on the second mixed number, you can subtract the whole number and the fractions separately. Remember to use the lowest common denominator on the fractions.

5 − 1 = 4

1/3 − 1/4 = 4/12 − 3/12 = 1/12

Therefore, the result is 4¹⁄₁₂.

37) The correct answer is: 20%

2 ÷ 10 = 0.20
0.20 = 20%

38) The correct answer is: 50

Reminder: For estimation problems like this, round the numbers up or down to the nearest whole number.

201 is rounded down to 200, and 3.9 is rounded up to 4. Then divide: 200 ÷ 4 = 50

39) 1.25 + .655 + .002 = ?

The correct answer is: 1.907

Remember to line up the decimal points as follows:

1.250
0.655
0.002
‾
1.907

40) The correct answer is: 20%

In order to calculate a discount, you must first determine how much the item was marked down.

$15 − $12 = $3

Then divide the mark down by the original price.

3 ÷ 15 = 0.20

Finally, convert the decimal to a percentage.

0.20 = 20%

41) The correct answer is: 9A

Final Score = A + B + C

B = 2A

C = 3B = 3 × 2A = 6A

Now express the original equation in terms of A:

A + B + C = A + 2A + 6A = 9A

42) The correct answer is: 0.15

Remember to do long division until you have no remainder.

```
        .15
   40) 6.00
       4.0
       ‾‾‾
       2.00
       2.00
       ‾‾‾‾
          0
```

43) The correct answer is: –2

Deal with the part of the equation inside the parentheses first.

$(-12 + 6) \div 3 =$
$-6 \div 3$

Then do the division.

$-6 \div 3 = -2$

44) The correct answer is: $x - 2$

For problems like this, look at the integers in the equation above. In this problem the integers are –4 and 2. We know that we have to divide –4 by 2 because the dividend is $(x + 2)$.

$-4 \div 2 = -2$

We also know that we have to divide x^2 by x, because these are the first terms in each set of parentheses: $x^2 \div x = x$

Now combine the two parts: $-2 + x = x - 2$

You can check your result as follows: $(x + 2)(x - 2) = x^2 - 2x + 2x - 4 = x^2 - 4$

45) The correct answer is: 2

To solve this type of problem, do multiplication of the items in parentheses first.

$5x - 2(x + 3) = 0$
$5x - 2x - 6 = 0$

Then deal with the integers by putting them on one side of the equation.

$5x - 2x - 6 + 6 = 0 + 6$
$3x = 6$

Then solve for x.

$3x = 6$
$x = 6 \div 3$
$x = 2$

46) The correct answer is: $x^2 + 6xy + 9y^2$

$(x + 3y)^2 = (x + 3y)(x + 3y)$

This type of algebraic expression is known as a polynomial. When multiplying polynomials, you should use the FOIL method.

This means that you multiply the terms two at a time from each of the two parts of the equation in this order:

First – Outside – Inside – Last

FIRST: Multiply the first term from the first set of parentheses with the first term from the second set of parentheses.

$x \times x = x^2$

OUTSIDE: Multiply the first term from the first set of parentheses with the second term from the second set of parentheses.

$x \times 3y = 3xy$

INSIDE: Multiply the second term from the first set of parentheses with the first term from the second set of parentheses.

$3y \times x = 3xy$

LAST: Multiply the second term from the first set of parentheses with the second term from the second set of parentheses.

$3y \times 3y = 9y^2$

Then we add all of the above parts together to get:

$x^2 + 3xy + 3xy + 9y^2 =$

$x^2 + 6xy + 9y^2$

47) The correct answer is: $x^2 + 2xy - 3y^2$

Remember to use the FOIL method when you multiply.

FIRST: $x \times x = x^2$

OUTSIDE: $x \times -y = -xy$

INSIDE: $3y \times x = 3xy$

LAST: $3y \times -y = -3y^2$

Then add all of the above once you have completed FOIL.

$x^2 - xy + 3xy - 3y^2 = x^2 + 2xy - 3y^2$

48) The correct answer is: 156

To solve this problem, put in the values for x and y and multiply. Remember to be careful when multiplying negative numbers.

$6x^2 - xy + y^2 =$
$(6 \times 5^2) - (5 \times -1) + (-1^2) =$
$(6 \times 5 \times 5) - (-5) + 1 =$
$(6 \times 25) + 5 + 1 =$
$150 + 5 + 1 =$
156

49) The correct answer is: (7.25A + 10.50B)

The two people are working at different per hour costs, so each person needs to have an individual variable.

A for the number of hours for the first person
B for the number of hours for the second person

So the equation for wages for the first person is: $(7.25 \times A)$
The equation for the wages for the second person is: $(10.50 \times B)$

The total cost of the wages for this job is the sum of the wages of these two people.

$(7.25 \times A) + (10.50 \times B) = (7.25A + 10.50B)$

50) The correct answer is: 8π

The circumference of a circle is always calculated by using this formula.

π times the diameter

(The diameter of a circle is always equal to the radius times 2.)

So, the diameter for this circle is 4 × 2 = 8. Therefore, the circumference is 8π.

51) The correct answer is: 8^{10}

This question tests your knowledge of exponent laws. First look to see whether your base number is the same on each part of the equation. (8 is the base number for each part of this equation.)

If the base number is the same, and the problem asks you to multiply, you simply add the exponents.

$8^7 \times 8^3 = 8^{(7+3)} = 8^{10}$

NOTE: If the base number is the same, and the problem asks you to *divide*, you *subtract* the exponents.

52) The correct answer is: 4

In order to solve inequalities, deal with the whole numbers on each side of the equation first.

$20 - \frac{3}{4}X > 17 =$
$(20 - 20) - \frac{3}{4}X > (17 - 20) =$
$-\frac{3}{4}X > -3$

Then deal with the fraction.

$-\frac{3}{4}X > -3 =$
$4 \times -\frac{3}{4}X > -3 \times 4 =$
$-3X > -12$

Then deal with the remaining whole numbers.

$-3X > -12 =$
$-3X \div 3 > -12 \div 3 =$
$-X > -4$

Then, deal with the negative number.

$-X > -4 =$
$-X + 4 > -4 + 4 =$
$-X + 4 > 0$

Finally, isolate the unknown variable as a positive number.

$-X + 4 > 0 =$
$-X + X + 4 > 0 + X =$
$4 > X =$
$X < 4$

53) The correct answer is: −24

Complete the operations inside the parentheses first.

Remember to be careful when multiplying the negative numbers.

$-6(4 - 1) - 2(5 - 2) =$

$-6(3) - 2(3) =$

$(-6 \times 3) - (2 \times 3) =$

$-18 - 6 =$

-24

54) The correct answer is: 3

Remember that when you see numbers between lines like this | −3 |, you are being asked the absolute value. Absolute value is always a positive number. So for this question:

| 3 − 6 | = | −3 |

| −3 | = 3

55) The correct answer is: 6.25×10^2

Scientific notation means that you have to give the number as a multiple of 10^2, in other words, as a factor of 100.

We know that 625 divided by 100 is 6.25.

So the answer is 6.25×10^2.

56) The correct answer is: $5^{½}$

This question is testing your knowledge of exponent laws. Remember that $\sqrt{x} = x^{½}$

57) The correct answer is: (−6, 0) and (0, 6)

To solve problems like this one, begin by substituting 0 for x.
y = x + 6
y = 0 + 6
y = 6

Therefore, the coordinates (0, 6) represent the y intercept.

Now substitute 0 for y.

y = x + 6
0 = x + 6
0 − 6 = x + 6 − 6
−6 = x

So, the coordinates (−6, 0) represent the x intercept.

58) The correct answer is: 8x − 2y

First perform the operations on the parentheses.

(5x + 7y) + (3x − 9y) =
5x + 7y + 3x − 9y

Then place the x and y terms together.

5x + 3x + 7y − 9y

Finally add or subtract the x and y terms.

$5x + 3x + 7y - 9y =$
$8x - 2y$

59) The correct answer is: $-x^2 + 8x - 12$

Remember to perform the operations on the parentheses first and to be careful with the negatives.

$(5x^2 + 3x - 4) - (6x^2 - 5x + 8) = 5x^2 + 3x - 4 - 6x^2 + 5x - 8$

Then place the x or y terms together.

$5x^2 - 6x^2 + 3x + 5x - 4 - 8$

Finally add or subtract the like terms.

$5x^2 - 6x^2 + 3x + 5x - 4 - 8 = -x^2 + 8x - 12$

60) The correct answer is: $3x^2 - 10x - 8$

Remember to use the FOIL method when you multiply.

FIRST: $x \times 3x = 3x^2$
OUTSIDE: $x \times 2 = 2x$
INSIDE: $-4 \times 3x = -12x$
LAST: $-4 \times 2 = -8$

Then add all of the above once you have completed FOIL.

$3x^2 + 2x + -12x + -8 =$
$3x^2 + 2x - 12x - 8 =$
$3x^2 - 10x - 8$

61) The correct answer is: $3\sqrt{7}$

In order to add square roots like this, you need to add the numbers in front of the square root sign.

$\sqrt{7} + 2\sqrt{7} =$
$1\sqrt{7} + 2\sqrt{7} =$
$3\sqrt{7}$

62) The correct answer is: $(x + 5)(x - 4)$

We know that for any problem like this, the answer will be in the format: $(x + ?)(x - ?)$

We know that we need to have a plus sign in one set of parentheses and a minus sign in the other set of parentheses because 20 is negative, and we can get a negative number in problems like this only if we multiply a negative and a positive.

We also know that the factors of 20 need to be one number different than each other because the middle term is x, in other words 1x. The only factors of twenty that meet this criterion are 4 and 5. Therefore the answer is $(x + 5)(x − 4)$

63) The correct answer is: 4.5

Perform the operations on the parentheses first.

$(−5 − (−14)) ÷ 2 =$
$(−5 + 14) ÷ 2 =$
$9 ÷ 2$

Then divide: $9 ÷ 2 = 4.5$

64) The correct answer is: .35A + .65B

The two tests are being given different percentages, so each assignment needs to have its own variable.

A for test A
B for test B

So the value of test A is .35A
The value of test B is .65B

The final grade is the sum of the values of these two variables: .35A + .65B

65) The correct answer is: $x^2 − 8xy + 16y^2$

$(x − 4y)^2 = (x − 4y)(x − 4y)$

This is another polynomial problem. When multiplying polynomials, you should use the FOIL method.

First − Outside − Inside − Last

FIRST: $x × x = x^2$
OUTSIDE: $x × −4y = −4xy$
INSIDE: $−4y × x = −4xy$
LAST: $− 4y × − 4y = 16y^2$

Then we add all of the above parts together to get: $x^2 − 8xy + 16y^2$

66) The correct answer is: 3

To solve this type of problem, do multiplication of the items in parentheses first.

$4x - 3(x + 2) = -3$
$4x - 3x - 6 = -3$

Then deal with the integers by putting them on one side of the equation.

$4x - 3x - 6 + 6 = -3 + 6$
$4x - 3x = 3$

Then solve for x.

$4x - 3x = 3$
$x = 3$

67) The correct answer is: (x + 3)

In order to solve this type of problem, you must do long division of the polynomial.

$$\begin{array}{r} x + 3 \\ x - 4 \overline{) x^2 - x - 12} \\ \underline{x^2 - 4x} \\ 3x - 12 \\ \underline{3x - 12} \\ 0 \end{array}$$

68) The correct answer is: –2

Deal with the part of the equation inside the parentheses first.

$(3 + -13) \div 5 =$
$-10 \div 5$

Then do the division.

$-10 \div 5 = -2$

69) The correct answer is: .20A + .25B

Remember that each item needs to have its own variable. A for apples and B for bananas. So the total value of the apples is .20A and the total value of the bananas is .25B

The total value of the purchase is the sum of the values of these two variables.

.20A + .25B

70) The correct answer is: 0.12π

The area of a circle is always: π times the radius squared.

Therefore, the area of circle A is: $0.4^2\pi = 0.16\pi$

The area of circle B is: $0.2^2\pi = 0.04\pi$

To calculate the difference in area between the two circles, we then subtract.

$0.16\pi - 0.04\pi = 0.12\pi$

71) The correct answer is: −18

To solve this problem, put in the values for x and y and multiply.

Tip: Remember to be careful when multiplying negative numbers.

$2x^2 + 3xy - y^2 =$
$(2 \times 3^2) + (3 \times 3 \times -3) - (-3^2) =$
$(2 \times 3 \times 3) + (3 \times 3 \times -3) - (-3 \times -3) =$
$(2 \times 9) + (3 \times -9) - (9) =$
$18 + (-27) - 9 =$
$18 - 27 - 9 =$
-18

72) The correct answer is: $3x^2 - 2xy - y^2$

FIRST: $x \times 3x = 3x^2$
OUTSIDE: $x \times y = xy$
INSIDE: $-y \times 3x = -3xy$
LAST: $-y \times y = -y^2$

Then add all of the above once you have completed FOIL.

$3x^2 + xy + -3xy + -y^2 =$
$3x^2 + xy - 3xy - y^2 =$
$3x^2 - 2xy - y^2$

73) The correct answer is: $\sqrt{6}$

In order to multiply two square roots, multiply the numbers inside the square roots.

$2 \times 3 = 6$

Then put this result inside a square root symbol for your answer: $\sqrt{6}$

74) The correct answer is: −15

Remember to complete the operations inside the parentheses first and to be careful when multiplying the negative numbers.

–3(5 – 2) – 6(4 – 3) =
–3(3) – 6(1) =
(–3 × 3) – (6 × 1) =
–9 – 6 =
–15

75) The correct answer is: $X < 5$

To solve inequalities like this one, deal with the whole numbers on each side of the equation first.

$20 - \frac{4}{5}X > 16 =$
$20 - 20 - \frac{4}{5}X > 16 - 20$
$-\frac{4}{5}X > -4$

Then deal with the fraction.

$-\frac{4}{5}X > -4 =$
$5 \times -\frac{4}{5}X > -4 \times 5 =$
$-4X > -20$

Then deal with the remaining whole numbers.

Remember that if you are multiplying or dividing by a negative number in any inequality problem, you have to reverse the direction of the inequality symbol.

$-4X > -20$
$-4X \div -4 > -20 \div -4$
$X < 5$

76) The correct answer is: –2

Deal with the part of the equation inside the parentheses first: $(-12 + 8) \div 2 = -4 \div 2$

Then do the division: $-4 \div 2 = -2$

77) The correct answer is: 1

Do multiplication of the items in parentheses first.

$7x - 5(x + 1) = -3$
$7x - 5x - 5 = -3$

Then deal with the integers by putting them on one side of the equation.

$7x - 5x - 5 + 5 = -3 + 5$
$7x - 5x = 2$

Then solve for x.

$2x = 2$
$x = 1$

78) The correct answer is: $2x^2 - 5xy + 2y^2$

Remember to use the FOIL method when you multiply.

FIRST: $x \times 2x = 2x^2$
OUTSIDE: $x \times -y = -xy$
INSIDE: $-2y \times 2x = -4xy$
LAST: $-2y \times -y = 2y^2$

Then add all of the above once you have completed FOIL.

$2x^2 + -xy + -4xy + 2y^2 =$
$2x^2 - xy - 4xy + 2y^2 =$
$2x^2 - 5xy + 2y^2$

79) The correct answer is: $2x^2 - 7xy + 3y^2$

FIRST: $2x \times x = 2x^2$
OUTSIDE: $2x \times -3y = -6xy$
INSIDE: $-y \times x = -xy$
LAST: $-y \times -3y = 3y^2$

Then add all of the above once you have completed FOIL.

$2x^2 + -6xy + -xy + 3y^2 =$
$2x^2 - 7xy + 3y^2$

80) The correct answer is: 20

Put in the values for x and y and multiply.

$3x^2 - xy + y^2 =$
$(3 \times 2^2) - (2 \times -2) + (-2^2) =$
$(3 \times 2 \times 2) - (2 \times -2) + (-2 \times -2) =$
$(3 \times 4) - (2 \times -2) + (4) =$
$12 - (-4) + 4 =$
$12 + 4 + 4 =$
20

81) The correct answer is: 12π

The circumference of a circle is always: $\pi \times$ diameter

The diameter for this circle is $6 \times 2 = 12$

Therefore, the circumference is 12π.

82) The correct answer is: 8

For inequalities, deal with the whole numbers on each side of the equation first.

$20 - \frac{1}{4}X > 18 =$
$(20 - 20) - \frac{1}{4}X > (18 - 20) =$
$-\frac{1}{4}X > -2$

Then deal with the fraction.

$-\frac{1}{4}X > -2 =$
$4 \times -\frac{1}{4}X > -2 \times 4 =$
$-X > -8$

Then, deal with the negative number.

$-X > -8 =$
$-X + 8 > -8 + 8 =$
$-X + 8 > 0$

Finally, isolate the unknown variable as a positive number.

$-X + 8 > 0 =$
$-X + X + 8 > 0 + X =$
$8 > X =$
$X < 8$

83) The correct answer is: –6

Remember to complete the operations inside the parentheses first and to be careful when multiplying the negative numbers.

$-5(3 - 1) - 2(5 - 7) =$

$-5(2) - 2(-2) =$

$(-5 \times 2) - (2 \times -2) =$

$-10 + 4 = -6$

84) The correct answer is: 3^7

Remember to add the exponents for multiplication problems like this one.

$3^4 \times 3^3 = 3^{3 + 4} = 3^7$

85) The correct answer is: $4x^2 + 20xy + 25y^2$

$(2x + 5y)^2 = (2x + 5y)(2x + 5y)$

FIRST: $2x \times 2x = 4x^2$
OUTSIDE: $2x \times 5y = 10xy$
INSIDE: $5y \times 2x = 10xy$
LAST: $5y \times 5y = 25y^2$

Then we add all of the above parts together to get: $4x^2 + 20xy + 25y^2$

86) The correct answer is: –10

Complete the subtraction inside the parentheses first. Remember to be careful when multiplying the negative numbers.

$-2(4 - 1) - 4(3 - 2) =$
$-2(3) - 4(1) = -6 - 4 =$
-10

87) The correct answer is: 6

$5x - 4(x + 2) = -2$

$5x - 4x - 8 = -2$

$x - 8 = -2$

$x - 8 + 8 = -2 + 8$

$x = 6$

88) The correct answer is: –3

Deal with the part of the equation inside the parentheses first.

$(-10 + 1) \div 3 = -9 \div 3$

Then do the division.

$-9 \div 3 = -3$

89) The correct answer is: 37

Put in the values for x and y and multiply.

$x^2 - xy + y^2 =$
$(4^2) - (4 \times -3) + (-3^2) =$
$(4 \times 4) - (4 \times -3) + (-3 \times -3) =$
$16 - (-12) + (9) =$
$16 + 12 + 9 = 37$

90) The correct answer is: 18π

Circumference = $\pi \times$ diameter

91) The correct answer is: 2

To solve this type of problem, do multiplication of the items in parentheses first.

$x - 2(x + 3) = -8$
$x - 2x - 6 = -8$

Then deal with the integers by putting them on one side of the equation.

$x - 2x - 6 + 6 = -8 + 6$
$x - 2x = -2$

Then solve for x.

$x - 2x = -2$
$-x = -2$
$x = 2$

92) The correct answer is: $x^2 - y^2$

FIRST: $x \times x = x^2$
OUTSIDE: $x \times y = xy$
INSIDE: $-y \times x = -xy$
LAST: $-y \times y = -y^2$

Then add all of the above once you have completed FOIL.

$x^2 + xy + -xy - y^2 = x^2 - y^2$

93) The correct answer is: $\sqrt{16}$

If you are asked to multiply two square roots, multiply the numbers inside the square roots: $8 \times 2 = 16$

Then put this result inside a square root symbol for your answer: $\sqrt{16}$

94) The correct answer is: $2xy(1 - 4x + 3xy)$

In order to factor an equation, you must figure out what variables are common to each term of the equation. Let's look at this equation:

$2xy - 8x^2y + 6y^2x^2$

We can see that each term contains x. We can also see that each term contains y. So, now let's factor out xy.

$2xy - 8x^2y + 6y^2x^2 =$
$xy(2 - 8x + 6xy)$

Then, think about integers. We can see that all of the terms inside the parentheses are divisible by 2. Now let's factor out the 2. In order to do this, we divide each term inside the parentheses by 2.

$xy(2 - 8x + 6xy) =$
$2xy(1 - 4x + 3xy)$

95) The correct answer is: $2x - 1$

This question is asking you to simplify the terms in the parentheses. First, you should look to see if there is any subtraction or if any of the numbers are negative. In this problem, the second part of the equation is subtracted. So we need to do the operation on the second set of parentheses first.

$(x + 3) - (4 - x) =$
$x + 3 - 4 + x$

Now simplify for the common terms.

$x + 3 - 4 + x =$
$x + x + 3 - 4 =$
$2x - 1$

96) The correct answer is: 32

For questions like this one, try to find the pattern of relationship between the numbers. Here, we can see that:

$2 \times 2 = 4$
$4 \times 2 = 8$
$8 \times 2 = 16$

In other words, the next number in the sequence is always double the previous number. Therefore the answer is: $16 \times 2 = 32$

97) The correct answer is: 5^6

This question tests your knowledge of exponent laws. First look to see whether your base number is the same on each part of the equation. (5 is the base number for each part of this equation.) If the base number is the same, and the problem asks you to divide, you simply subtract the exponents.

$5^8 \div 5^2 = 5^{8-2} = 5^6$

98) The correct answer is: $3x^2 - 15x - 3$

For simplification problems, you should look to see if there is any subtraction or if any of the numbers are negative. In this problem, the second part of the equation is subtracted. So we need to perform the operation on the second set of parentheses first.

$(4x^2 - 5x - 3) - (x^2 + 10x) =$
$(4x^2 - 5x - 3) - x^2 - 10x$

Then we can remove the remaining parentheses.

$(4x^2 - 5x - 3) - x^2 - 10x =$
$4x^2 - 5x - 3 - x^2 - 10x$

Now simplify for the integers and common variables.

$4x^2 - 5x - 3 - x^2 - 10x =$
$4x^2 - x^2 - 5x - 10x - 3 =$
$3x^2 - 15x - 3$

99) The correct answer is: 4 hours

Remember to read questions like this one very carefully. If the car travels at 60 miles an hour and needs to go 240 more miles, we need to divide the miles to travel by the miles per hour.

miles to travel ÷ miles per hour = time remaining

So, if we substitute the values from the question, we get:

$240 \div 60 = 4$

In other words, the total time is 4 hours.

100) The correct answer is: 60

Permutations are like combinations, except permutations take into account the order of the items in each group. In order to calculate the number of permutations of size S taken from N items, you should use this formula:

$N! \div (N - S)!$

For the question above: $N = 5$ and $S = 3$

$N! \div (N - S)! =$
$(5 \times 4 \times 3 \times 2 \times 1) \div (5 - 3)! =$
$(5 \times 4 \times 3 \times 2) \div 2 =$
$120 \div 2 =$
60

101) The correct answer is: $y > 0$

This is an inequality problem. Notice that both equations contain $x - 1$.

Therefore, we can substitute y for $x - 1$ in the first equation:

$x - 1 > 0$
$x - 1 = y$
$y > 0$

102) The correct answer is: $jn - mk$

In order to find the determinant for a two-by-two matrix, you need to cross multiply and then subtract. So j is multiplied by n and m is multiplied by k. Then we subtract the two terms to get the determinant: $jn - mk$

103) The correct answer is: (−1, −1)

This question covers coordinate geometry. Remember that in order to find midpoints on a line, you need to use the following formula:

For two points on a graph (x_1, y_1) and (x_2, y_2), the midpoint is:

$(x_1 + x_2) \div 2$, $(y_1 + y_2) \div 2$

Now calculate for x and y.

$(-5 + 3) \div 2$ = midpoint x, $(3 + -5) \div 2$ = midpoint y
$-2 \div 2$ = midpoint x, $-2 \div 2$ = midpoint y
-1 = midpoint x, -1 = midpoint y

104) The correct answer is: 5 pairs

Let's say that the number of pairs of socks is S and the number of pairs of shoes is H.

Now let's make an equation to express the above problem.

$(S \times \$2) + (H \times \$25) = \$85$

We know that the number of pairs of shoes is 3, so let's put that in the equation and solve it.

$(S \times \$2) + (H \times \$25) = \$85$
$(S \times \$2) + (3 \times \$25) = \$85$
$(S \times \$2) + \$75 = \$85$
$(S \times \$2) + 75 - 75 = \$85 - \$75$
$(S \times \$2) = \$10$
$\$2S = \$10$
$\$2S \div 2 = \$10 \div 2$
$S = 5$

105) The correct answer is: 3

In order to calculate the slope of a line, you need this formula:

$y = mx + b$

NOTE: *m* is the slope and *b* is the *y* intercept (the point at which the line crosses the *y* axis).

Now solve for the numbers given in the problem.

$y = mx + b$
$14 = m3 + 5$
$14 - 5 = m3 + 5 - 5$
$9 = m3$
$9 \div 3 = m$
$3 = m$

106) The correct answer is: 2

In equations that have both integers and square roots, first deal with the integers that are outside the parentheses.

$5 + 5(3\sqrt{x} + 4) = 55$
$5 + 15\sqrt{x} + 20 = 55$
$25 + 15\sqrt{x} = 55$
$25 - 25 + 15\sqrt{x} = 55 - 25$
$15\sqrt{x} = 30$

Then divide.

$15\sqrt{x} = 30$
$(15\sqrt{x}) \div 15 = 30 \div 15$
$\sqrt{x} = 2$

107) The correct answer is: (3, 8)

For questions on systems of equations like this one, you should look at the multiplication equation first. Ask yourself, what are the factors of 24?

We know that 24 is the product of the following:

$1 \times 24 = 24$
$2 \times 12 = 24$
$3 \times 8 = 24$
$4 \times 6 = 24$

Now add each of the two factors together to solve the first equation.

1 + 24 = 25

2 + 12 = 14

3 + 8 = 11

4 + 6 = 10

(3, 8) solves both equations. Therefore, it is the correct answer.

108) The correct answer is: *a* must be equal to *b* and *xi* must be equal to *yi*

Two complex numbers are equal if and only if their real parts are equal and their imaginary parts are equal.

109) The correct answer is: $\sqrt{13}$

The length of the hypotenuse is always the square root of the sum of the squares of the other two sides of the triangle.

hypotenuse length L = $\sqrt{M^2 + N^2}$

Now put in the values for the above problem.

L = $\sqrt{M^2 + N^2}$
L = $\sqrt{3^2 + 2^2}$
L = $\sqrt{9 + 4}$
L = $\sqrt{13}$

110) The correct answer is: 5 = $\log_2 32$

Logarithmic functions are just another way of expressing exponents. Remember that:

y^x = Z is always the same as x = $\log_y Z$

111) The correct answer is: 1 ÷ x^7

Remember that a negative exponent is always equal to 1 divided by the variable with a positive exponent.

Therefore, x^{-7} = 1 ÷ x^7

112) The correct answer is: –2 + i

To solve this type of problem, do the operations on the parentheses first.

(2 – 2i) – (4 – 3i) = 2 – 2i – 4 + 3i

Then group the real and imaginary numbers together.

2 – 2i – 4 + 3i =
2 – 4 – 2i + 3i =
–2 + i

113) The correct answer is: $\dfrac{35x}{4}$

To solve this problem, multiply the numerator of the first fraction by the numerator of the second fraction to calculate the numerator of the new fraction. Then multiply the denominators in order to get the new denominator. Then simplify, if possible.

$$\dfrac{5x^3}{4} \times \dfrac{7}{x^2} = \dfrac{35x^3}{4x^2} = \dfrac{35x}{4}$$

114) The correct answer is: $18x^5$

Remember to multiply the base numbers, add the exponents, and be careful with the negatives.

$$(-3x)(-6x^4) =$$

–3x¹ × –6⁴ =

$18x^5$

115) The correct answer is: $\dfrac{x^2 + 2x}{x^2 + 9x + 20}$

Factor the numerators and denominators, then cancel out and re-simplify.

$$\dfrac{x^2 + 8x + 12}{x^2 + 8x + 16} \times \dfrac{x^2 + 4x}{x^2 + 11x + 30} =$$

$$\dfrac{(x+2)(x+6)}{(x+4)(x+4)} \times \dfrac{x(x+4)}{(x+5)(x+6)} =$$

$$\dfrac{x(x+2)}{(x+4)(x+5)} =$$

$$\frac{x^2+2x}{x^2+9x+20}$$

116) The correct answer is: $\dfrac{8x+5}{x^3+2x^2+x}$

Factor the denominators of each fraction in order to help you find the lowest common denominator (LCD). Then re-simplify after you have determined the LCD.

$$\frac{3}{x^2+2x+1}+\frac{5}{x^2+x}=$$

$$\frac{3}{(x+1)(x+1)}+\frac{5}{x(x+1)}=$$

$$\frac{x}{x}\times\frac{3}{(x+1)(x+1)}+\frac{5}{x(x+1)}\times\frac{(x+1)}{(x+1)}=$$

$$\frac{3x}{x(x+1)(x+1)}+\frac{5x+5}{x(x+1)(x+1)}=$$

$$\frac{8x+5}{x(x+1)(x+1)}=$$

$$\frac{8x+5}{x(x^2+2x+1)}$$

$$\frac{8x+5}{x^3+2x^2+x}$$

117) The correct answer is: −24

Eliminate the fraction by multiplying both sides of the equation by −8.

$$3=-\frac{1}{8}x$$

$$3\times-8=-\frac{1}{8}x\times-8$$

−24 = x

118) The correct answer is: 10

Square both sides of the equation and then isolate x in order to solve the problem.

$\sqrt{4x-4} = 6$

$\sqrt{4x-4}^2 = 6^2$

4x − 4 = 36

4x − 4 + 4 = 36 + 4

4x = 40

x = 10

119) The correct answer is: y = 2/5x − 6/5

Find the slope, represented by variable *m*, by putting the stated values into the slope formula.

$\dfrac{y_2 - y_1}{x_2 - x_1} = m$

$\dfrac{2-0}{8-3} = \dfrac{2}{5}$

Then calculate the y intercept, represented by variable *b*, by putting the values for x, y, and *m* into the slope-intercept formula.

y = mx − b

0 = (2/5 × 3) − b

6/5 = b

Finally, express as the slope-intercept equation, using variables x and y.

y = 2/5x − 6/5

120) The correct answer is: 16

This question tests your knowledge of order of operations. Remember to do operations on parentheses first, if any. Then do the exponents, if any. Next, do the multiplication and division (from left to right), and finally the addition and subtraction.

50 ÷ 5 + 36 ÷ 6 =

(50 ÷ 5) + (36 ÷ 6) =

10 + 6 = 16

121) The correct answer is: $b = 24/5$ and $m = -3/5$

Plug in 0 for x in order to calculate b, the y intercept.

$3x + 5y = 24$

$3 \times 0 + 5y = 24$

$5y = 24$

$y = 24/5$

(0, 24/5) are the coordinates for the y intercept.

Now put in 0 for y in order to calculate x.

$3x + 5y = 24$

$3x + 5 \times 0 = 24$

$3x = 24$

x = 8

(8, 0) are the coordinates for the x intercept.

Next use the slope formula to calculate the slope. Remember to simplify the fraction as much as possible.

$$\frac{y_2 - y_1}{x_2 - x_1} = m$$

$$\frac{0 - 24/5}{8 - 0} = m$$

$$-\frac{24/5}{8} = m$$

$$-24/5 \div 8 = m$$

$$-24/5 \times 1/8 = m$$

$$-24/40 = m$$

$$-\frac{3 \times 8}{5 \times 8} = m$$

$-3/5 = m$

122) The correct answer is: $\dfrac{7x}{4}$

Invert and multiply by the second fraction. Cancel out, if possible. Then simplify the resulting fraction in order to get your final result.

$$\frac{7x+7}{x} \div \frac{4x+4}{x^2} =$$

$$\frac{7x+7}{x} \times \frac{x^2}{4x+4} =$$

$$\frac{7(x+1)}{x} \times \frac{x^2}{4(x+1)} =$$

$$\frac{7x^2}{4x} =$$

$$\frac{7x}{4}$$

123) The correct answer is: $7x^7$

This is another exponent problem in a slightly different form. Again, you have to remember the basic principles of factoring and the basic principles of multiplying numbers that have exponents. Remember to multiply base numbers and add exponents.

$$49x^8 = 7x(\ )$$

$$7x \times 7x^7 = 49x^8$$

124) The correct answer is: $-x^8 + 11x^5$

Deal with the negative sign in front of the second set of parentheses. Then group like terms together in order to simplify.

$$(4x^8 + 5x^5 - 7) - (-6x^5 + 5x^8 - 7) =$$

$$4x^8 + 5x^5 - 7 + 6x^5 - 5x^8 + 7 =$$

$$4x^8 - 5x^8 + 5x^5 + 6x^5 - 7 + 7 =$$

$$-x^8 + 11x^5$$

125) The correct answer is: $6x$

Compare the denominator of the first fraction with the denominator of the second fraction and divide, if possible, in order to find the common factor: $9x^2 \div 3x = 3x$

Now multiply the numerator on the first fraction by this result in order to get the numerator of the second fraction: $2 \times 3x = 6x$

126) The correct answer is: $\dfrac{11x + 16}{x^2 + 2x}$

The LCD in this problem is $x^2 + 2x$. Remember to multiply the numerator and denominator by the same amounts when converting to the LCD.

$$\frac{8}{x} + \frac{3}{x+2} =$$

$$\frac{8}{x} \times \frac{x+2}{x+2} + \frac{3}{x+2} \times \frac{x}{x} =$$

$$\frac{8x + 16}{x^2 + 2x} + \frac{3x}{x^2 + 2x} =$$

$$\frac{8x+16+3x}{x^2+2x} =$$

$$\frac{11x+16}{x^2+2x}$$

127) The correct answer is: $\frac{3B}{C}$

Isolate D by eliminating the fraction and dividing by C.

$$B = \frac{1}{3}CD$$

$$B \times 3 = \frac{1}{3} \times 3CD$$

$$3B = CD$$

$$3B \div C = CD \div C$$

$$\frac{3B}{C} = D$$

128) The correct answer is: –2/3 and –2

Factor the equation and then substitute 0 in each part of the factored equation to get your result.

$$6x^2 + 16x + 8 = 0$$

$(3x + 2)(2x + 4) = 0$

Now substitute 0 for x in the first pair of parentheses.

$(3 \times 0 + 2)(2x + 4) = 0$

$2(2x + 4) = 0$

$4x + 8 = 0$

$x = -2$

Then substitute 0 for x in the second pair of parentheses.

$(3x + 2)(2x + 4) = 0$

$(3x + 2)(2 \times 0 + 4) = 0$

$(3x + 2)4 = 0$

$12x + 8 = 0$

$12x + 8 - 8 = 0 - 8$

$12x = -8$

$x = -8/12$

$x = -2/3$

129) The correct answer is: x^{10}

$$\frac{x^2}{x^{-8}} = x^2 \div x^{-8} = x^{2--8} = x^{10}$$

130) The correct answer is: $\dfrac{XY}{W}$

Multiply each side of the equation by Z. Then divide by W in order to isolate Z.

$$W = \frac{XY}{Z}$$

$$W \times Z = \frac{XY}{Z} \times Z$$

WZ = XY

WZ ÷ W = XY ÷ W

$$Z = \frac{XY}{W}$$

131) The correct answer is: 904.32

The formula for the volume (V) of a cylinder is: $V = \pi r^2 h$

In other words, to calculate the volume of a cylinder you take π times the radius squared times the height.

Place the stated values into the equation in order to solve the problem.

V = $\pi r^2 h$

V = 3.14 × 4² × 18

V = 3.14 × 16 × 18

V = 904.32

132) The correct answer is: $\dfrac{14x - 20}{105x^2}$

You will know by now that you need to find the LCD and then perform the operation.

$$\dfrac{2}{15x} - \dfrac{4}{21x^2} =$$

$$\dfrac{2}{15x} \times \dfrac{7x}{7x} - \dfrac{4}{21x^2} \times \dfrac{5}{5} =$$

$$\dfrac{14x}{105x^2} - \dfrac{20}{105x^2} =$$

$$\dfrac{14x - 20}{105x^2}$$

133) The correct answer is: 17

Place the stated values into the equation and perform the operations in order to solve the problem.

$$A = \dfrac{1}{2}(B + C)d$$

$$120 = \dfrac{1}{2}(13 + C)8$$

$$120 \div 8 = \dfrac{1}{2}(13 + C)8 \div 8$$

$$15 = \frac{1}{2}(13+C)$$

$$15 \times 2 = \frac{1}{2} \times 2(13+C)$$

$$30 = 13 + C$$

$$17 = C$$

134) The correct answer is: $50a^2b^{12} - 40ab^8 - 100a^2b^5$

Step 1: Apply the distributive property of multiplication by multiplying the item in front of the opening parenthesis by each item inside the pair of parentheses.

Step 2: Add up the individual products in order to solve the problem.

$$10ab^5(5ab^7 - 4b^3 - 10a) =$$

$$(10ab^5 \times 5ab^7) - (10ab^5 \times 4b^3) - (10ab^5 \times 10a) =$$

$50a^2b^{12} - 40ab^8 - 100a^2b^5$

135) The correct answer is: –5

Assign a variable to the mystery number. In this case, we will call the number x. Then make an equation based on the information stated in the problem.

twice a number = $2x$

8 less than the number = $x - 8$

the sum of twice a number and 8 less than the number = $2x + x - 8$

the difference between –28 and the number = $-28 - x$

So the equation is: $2x + x - 8 = -28 - x$

Finally, solve the equation for x.

$2x + x - 8 = -28 - x$

$2x + x - 8 + 8 = -28 + 8 - x$

$2x + x = -20 - x$

$2x + x + x = -20 - x + x$

$4x = -20$

$x = -5$

136) The correct answer is: $z + 2$

Remember to invert and multiply. Then factor and re-simplify, cancelling out where needed.

$$\frac{z^2 + 7z + 10}{z^2 + 13z + 40} \div \frac{z + 8}{z^2 + 16z + 64} =$$

$$\frac{z^2 + 7z + 10}{z^2 + 13z + 40} \times \frac{z^2 + 16z + 64}{z + 8} =$$

$$\frac{(z+2)(z+5)}{(z+8)(z+5)} \times \frac{(z+8)(z+8)}{z+8} =$$

$$\frac{(z+2)}{1} = z + 2$$

137) The correct answer is: $x^2 + 1$

When you see fractions that have fractions within themselves, remember to treat the denominator as the division sign, and then invert and multiply the fractions as usual.

$$\frac{x + \frac{1}{x}}{\frac{1}{x}} =$$

$$\left(x + \frac{1}{x}\right) \div \frac{1}{x} =$$

$$\left(x + \frac{1}{x}\right) \times x =$$

$$x^2 + \frac{x}{x} =$$

$x^2 + 1$

138) The correct answer is: 10

Eliminate the integer, then the fraction, and then isolate *a* in order to solve the problem.

$$\frac{3a}{10} + 9 = 12$$

$$\frac{3a}{10} + 9 - 9 = 12 - 9$$

$$\frac{3a}{10} = 3$$

$$\frac{3a}{10} \times 10 = 3 \times 10$$

$3a = 30$

$a = 10$

139) The correct answer is: $y = -x + 5$

First you need to calculate slope (which is variable *m* in the slope-intercept equation) using the slope formula: $\dfrac{y_2 - y_1}{x_2 - x_1}$

Substitute the values for *x* and *y* from the table in order to calculate the slope.

$$\frac{y_2 - y_1}{x_2 - x_1} =$$

$$\frac{1 - 5}{4 - 0} =$$

$$\frac{-4}{4} = -1$$

We know from the information provided in the table that the *y* intercept (which is variable *b* in the slope-intercept equation) is 5 because of the coordinates (0, 5).

So we place these values into the slope-intercept formula in order to solve the problem.

$y = mx + b$

$y = -1x + 5$

$y = -x + 5$

140) The correct answer is: 9

Eliminate the fraction and the integer. Then eliminate the radical by squaring both sides of the equation. Finally, isolate x to solve the problem.

$$\frac{20}{\sqrt{x^2 + 7}} = 5$$

$$\frac{20}{\sqrt{x^2 + 7}} \times \sqrt{x^2 + 7} = 5 \times \sqrt{x^2 + 7}$$

$$20 = 5\sqrt{x^2 + 7}$$

$$20 \div 5 = (5\sqrt{x^2 + 7}) \div 5$$

$$4 = \sqrt{x^2 + 7}$$

$$4^2 = (\sqrt{x^2 + 7})^2$$

$16 = x^2 + 7$

$16 - 7 = x^2 + 7 - 7$

$9 = x^2$

Tip: Read these types of problems carefully. Sometimes they will ask you to solve for x^2 and other times they will ask you to solve for x.

141) The correct answer is: $4/5$

You have to find the cube root of the numerator and denominator in order to eliminate the radical.

$$\sqrt[3]{\frac{64}{125}} = \sqrt[3]{\frac{4 \times 4 \times 4}{5 \times 5 \times 5}} = \frac{4}{5}$$

142) The correct answer is: 5

Tip: To solve this problem, you need the distance formula.

$$d = \sqrt{(x_2 - x_1)^2 + (y_2 - y_1)^2}$$

$$d = \sqrt{(6-3)^2 + (4-0)^2}$$

$$d = \sqrt{3^2 + 16}$$

$$d = \sqrt{9 + 16}$$

$$d = \sqrt{25}$$

$$d = 5$$

143) The correct answer is: $y = -2x + 5$

Tip: Two lines are perpendicular if the product of their slopes is equal to –1.

Step 1: Calculate the slope. We can see that the slope of the line stated in the problem is ½. Because the lines are perpendicular, we calculate the slope of the new line with this formula: ½ × m = –1

So the slope of the perpendicular line is –2.

Step 2: To solve the problem, put the slope that you calculated in step 1 into the formula given in the problem.

For the given line: $y = \frac{1}{2}x + 5$

For the perpendicular line: $y = -2x + 5$

144) The correct answer is: 14

First you need to eliminate the fraction and simplify the result as far as possible. Then remove the common terms and integers in order to isolate *a* and solve the problem.

$$\frac{b^2 - ab + 24}{b - 12} = b - 2$$

$$\frac{b^2 - ab + 24}{b - 12} \times (b - 12) = (b - 2)(b - 12)$$

$$b^2 - ab + 24 = (b - 2)(b - 12)$$

$$b^2 - ab + 24 = b^2 - 14b + 24$$

$$b^2 - b^2 - ab + 24 - 24 = b^2 - b^2 - 14b + 24 - 24$$

$$-ab = -14b$$

$$a = 14$$

145) The correct answer is: $\dfrac{1}{25}$

Tip 1: When you see a fraction as an exponent, remember that you need to place the base number inside the radical sign. The denominator of the exponent is the nth root of the radical, and the numerator of the fraction becomes the new exponent. Here is an example: $x^{3/4} = (\sqrt[4]{x})^3$ *Tip 2:* When you see a negative exponent, you remove the negative sign on the exponent by expressing the number as a fraction, with 1 as the numerator. Here is an example: $x^{-6} = \dfrac{1}{x^6}$ So you need to combine these two principles in order to solve the problem.

$$125^{-2/3} = \frac{1}{125^{2/3}} = \frac{1}{\sqrt[3]{125}^2} = \frac{1}{(\sqrt[3]{5 \cdot 5 \cdot 5})^2} = \frac{1}{5^2} = \frac{1}{25}$$

146) The correct answer is: $x = \sqrt{5}$

First you need to eliminate the denominator of the fraction.

$$\frac{18}{\sqrt{x^2 + 4}} = 6$$

$$\frac{18}{\sqrt{x^2 + 4}} \times (\sqrt{x^2 + 4}) = 6 \times (\sqrt{x^2 + 4})$$

$18 = 6\sqrt{x^2 + 4}$

Then eliminate the integer in front of the radical.

$18 = 6\sqrt{x^2 + 4}$

$18 \div 6 = (6\sqrt{x^2 + 4}) \div 6$

$3 = \sqrt{x^2 + 4}$

Then square both sides of the equation in order to solve the problem.

$3 = \sqrt{x^2 + 4}$

$3^2 = (\sqrt{x^2 + 4})^2$

$9 = x^2 + 4$

$9 - 4 = x^2 + 4 - 4$

$5 = x^2$

$x = \sqrt{5}$

147) The correct answer is: $y = -0.5x + b$

Tip 1: If two lines are parallel, they will have the same slope. So we can use the put the same value for *m* into both equations. *Tip 2:* Note that the parallel lines will have a different *y* intercept.

148) The correct answer is: $\dfrac{4\sqrt{3}}{3}$

"Rationalize" means to remove the square root symbol by performing the necessary mathematical operations.

We remove the square root from the denominator as follows:

$\sqrt{\dfrac{16}{3}} = \dfrac{\sqrt{16}}{\sqrt{3}} = \dfrac{\sqrt{4 \times 4}}{\sqrt{3}} = \dfrac{4}{\sqrt{3}} = \dfrac{4 \times \sqrt{3}}{\sqrt{3} \times \sqrt{3}} = \dfrac{4\sqrt{3}}{3}$

149) The correct answer is: $106 - 19\sqrt{10}$

Tip 1: Don't panic when you see the radicals. This is just another type of FOIL problem.

Tip 2: When you multiply radicals, multiply the numbers in front of the radicals and then the numbers inside the radicals. Here is an example: $3\sqrt{3} \times 4\sqrt{2} = 12\sqrt{6}$

Now here is the solution to the problem.

$(\sqrt{2} - 5\sqrt{5})(3\sqrt{2} - 4\sqrt{5}) =$

$(\sqrt{2} \times 3\sqrt{2}) + (\sqrt{2} \times -4\sqrt{5}) + (-5\sqrt{5} \times 3\sqrt{2}) + (-5\sqrt{5} \times -4\sqrt{5}) =$

$(3 \times 2) + (-4\sqrt{10}) + (-15\sqrt{10}) + (20 \times 5) =$

$6 - 4\sqrt{10} - 15\sqrt{10} + 100 =$

$106 - 19\sqrt{10}$

150) The correct answer is: $4x^2y$

When the denominator of a fraction contains another fraction, treat the main fraction as the division sign. Then invert and multiply as usual.

$$\frac{4x}{\frac{1}{xy}} = 4x \div \frac{1}{xy} = 4x \times xy = 4x^2y$$

151) $64^{3/2} = ?$

The correct answer is: 512

Here is some further practice with some concepts we have seen earlier. Remember that when you see a fraction as an exponent, you need to place the base number inside the radical sign. The denominator of the exponent is the nth root of the radical, and the numerator of the fraction becomes the new exponent. Here is an example: $x^{3/4} = (\sqrt[4]{x})^3$

You will need to simplify the radical as much as possible.

So for our problem: $64^{3/2} = \sqrt{64}^3 = (\sqrt{8 \times 8})^3 = 8^3 = 512$

152) Simplify: $\dfrac{\sqrt{75}}{3} + \dfrac{5\sqrt{5}}{6}$

The correct answer is: $\dfrac{10\sqrt{3} + 5\sqrt{5}}{6}$

Find the LCD and then perform the operations, including simplification of the radical, in order to solve the problem.

$$\dfrac{\sqrt{75}}{3} + \dfrac{5\sqrt{5}}{6} =$$

$$\dfrac{\sqrt{75}}{3} \times \dfrac{2}{2} + \dfrac{5\sqrt{5}}{6} =$$

$$\dfrac{2\sqrt{75}}{6} + \dfrac{5\sqrt{5}}{6} =$$

$$\dfrac{2\sqrt{75} + 5\sqrt{5}}{6} =$$

$$\dfrac{2\sqrt{25 \times 3} + 5\sqrt{5}}{6} =$$

$$\dfrac{2 \times 5\sqrt{3} + 5\sqrt{5}}{6} =$$

$$\dfrac{10\sqrt{3} + 5\sqrt{5}}{6}$$

153) The correct answer is: $2 + \dfrac{5\sqrt{5}}{9}$

Here it appears that we have a mixed number on the second fraction. However, don't let this confuse you. The basic concepts are the same as in the preceding problem.

$$\dfrac{\sqrt{36}}{3} + 5\dfrac{\sqrt{5}}{9} =$$

$$\frac{\sqrt{36}}{3} + \frac{5\sqrt{5}}{9} =$$

$$\frac{\sqrt{36}}{3} \times \frac{3}{3} + \frac{5\sqrt{5}}{9} =$$

$$\frac{3\sqrt{36}}{9} + \frac{5\sqrt{5}}{9} =$$

$$\frac{3 \times 6}{9} + \frac{5\sqrt{5}}{9} =$$

$$\frac{18}{9} + \frac{5\sqrt{5}}{9} =$$

$$\frac{18 + 5\sqrt{5}}{9} =$$

$$2 + \frac{5\sqrt{5}}{9}$$

154) The correct answer is: $3\sqrt{2} + 35\sqrt{3}$

Find the squared factors of the amounts inside the radical signs. Then simplify.

$$\sqrt{18} + 4\sqrt{75} + 5\sqrt{27} =$$

$$\sqrt{2 \times 9} + 4\sqrt{3 \times 25} + 5\sqrt{3 \times 9} =$$

$$3\sqrt{2} + (4 \times 5)\sqrt{3} + (5 \times 3)\sqrt{3} =$$

$$3\sqrt{2} + 20\sqrt{3} + 15\sqrt{3} =$$

$$3\sqrt{2} + 35\sqrt{3}$$

155) The correct answer: 4 hot dogs

The number of hot dogs is *D* and the number of hamburgers is *H*.

Here is the equation to express the problem: $(D \times \$2.50) + (H \times \$4) = \$22$

We know that the number of hamburgers is 3, so put that in the equation and solve it.

$(D \times \$2.50) + (H \times \$4) = \$22$
$(D \times \$2.50) + (3 \times \$4) = \$22$
$(D \times \$2.50) + 12 = \$22$
$(D \times \$2.50) + 12 - 12 = \$22 - 12$
$(D \times \$2.50) = \$10$
$\$2.50D = \$10$
$\$2.50D \div \$2.50 = \$10 \div \$2.50$
$D = 4$

156) The correct answer is: $24x^4 + 18x^3 - 2x^2 - 24x - 40$

Use the distributive property of multiplication, group like terms together, and then simplify.

$(4x^2 + 3x + 5)(6x^2 - 8) =$

$(4x^2 \times 6x^2) + (3x \times 6x^2) + (5 \times 6x^2) + [(4x^2 \times -8) + (3x \times -8) + (5 \times -8)] =$

$24x^4 + 18x^3 + 30x^2 + (-32x^2 + -24x + -40) =$

$24x^4 + 18x^3 + 30x^2 - 32x^2 - 24x - 40 =$

$24x^4 + 18x^3 - 2x^2 - 24x - 40$

157) The correct answer is: 13^8

If the base number is the same, and the problem asks you to multiply, you add the exponents:

$13^3 \times 13^5 = 13^{3+5} = 13^8$

158) The correct answer is: $6xy(1 - 2x - 4xy)$

In order to factor an equation, you must figure out what terms are common to each term of the equation. Let's factor out xy.

$6xy - 12x^2y - 24y^2x^2 =$
$xy(6 - 12x - 24xy)$

Then, think about integers. We can see that all of the terms inside the parentheses are divisible by 6. Now let's factor out the 6. In order to do this, we divide each term inside the parentheses by 6.

$xy(6 - 12x - 24xy) =$
$6xy(1 - 2x - 4xy)$

159) The correct answer is: 15

To solve inequalities like this one, you should first solve the equation for x.

x − 5 < 0 =
x − 5 + 5 < 0 + 5=
x < 5

Now solve for y by replacing x with its value.

y < x + 10 =
y < 5 + 10 =
y < 15

160) The correct answer is: $2x^6\sqrt{6}$

Tip: When the two radicals symbols are together like this, you need to multiply them.

$$\sqrt{4x^8}\sqrt{6x^4} =$$

$$\sqrt{4x^8} \times \sqrt{6x^4} =$$

$$\sqrt{24x^{12}} = \sqrt{4 \times 6} \times \sqrt{x^{12}} = 2\sqrt{6} \times x^{\frac{12}{2}} = 2x^6\sqrt{6}$$

161) The correct answer is: 35

First, find the relationship between each of the numbers given. After looking at the numbers given above, we can see that:

7 + 7 = 14
14 + 7 = 21
21 + 7 = 28

Therefore, we have to add 7 to 28 in order to find the solution.

28 + 7 = 35

162) The correct answer is: (3, 0) and (0, 2)

Remember that for questions about x and y intercepts, you need to substitute 0 for x and y to solve the problem.

Solution for y intercept:

$4x^2 + 9y^2 = 36$
$4(0^2) + 9y^2 = 36$
$0 + 9y^2 = 36$
$9y^2 \div 9 = 36 \div 9$

$y^2 = 4$
$y = 2$

So the y intercept is $(0, 2)$

Solution for x intercept:

$4x^2 + 9y^2 = 36$
$4x^2 + 9(0^2) = 36$
$4x^2 + 0 = 36$
$4x^2 \div 4 = 36 \div 4$
$x^2 = 9$
$x = 3$

So the x intercept is $(3, 0)$

163) The correct answer is: $(3, -2)$

For two points on a graph (x_1, y_1) and (x_2, y_2), the midpoint is: $(x_1 + x_2) \div 2$, $(y_1 + y_2) \div 2$

Now calculate for x and y.

$(2 + 4) \div 2 =$ midpoint x, $(2 - 6) \div 2 =$ midpoint y
$6 \div 2 =$ midpoint x, $-4 \div 2 =$ midpoint y
$3 =$ midpoint x, $-2 =$ midpoint y

164) The correct answer is: -7

When you see numbers between lines like this, you need to determine the absolute value. Remember that the absolute value is always a positive number.

$- | 10 - 17 | =$
$- | -7 |$

So the absolute value of -7 is 7. But notice the negative sign in front of the absolute value symbol. Therefore, you finally have to give the negative of the absolute value.

$- | -7 | =$
$- (7) =$
-7

165) The correct answer is: $3i$

Note that it is not possible to find the square root of a negative number by using real numbers. Therefore, you will have to use imaginary numbers to solve this problem. Imaginary numbers are represented by the variable i.

So first determine what the square root of the number would be if the number were positive.

$\sqrt{9} = 3$

Now multiply that result by i.

3 × i = 3i

166) The correct answer is: –5

Tip: To find the determinant for a two-by-two matrix, you need to cross multiply and then subtract.

$$\begin{bmatrix} 4 & -1 \\ 3 & -2 \end{bmatrix}$$

So 4 is multiplied by –2 and 3 is multiplied by –1.

Then we subtract the two terms to get the determinant.

(4 × –2) – (3 × –1) =
–8 – (–3) =
–8 + 3 = –5

167) The correct answer is: $\log_3 243 = 5$

Logarithmic functions are just another way of expressing exponents. Remember that:

$\log_y Z = x$ is always the same as $y^x = Z$

So $3^5 = 243$ is the same as $\log_3 243 = 5$

168) The correct answer is: 10

To determine the number of combinations of S at a time that can be made from a set containing N items, you need this formula: $(N!) \div [(N - S)! \times S!]$

In the problem above, S = 2 and N = 5 (because there are five letters in the set).

Now substitute the values for S and N.

(5 × 4 × 3 × 2 × 1) ÷ [(5 – 2)! × (2 !)] =
(5 × 4 × 3 × 2) ÷ [(3 × 2 × 1) × (2 × 1)] =
120 ÷ 12 = 10

169) The correct answer is: (1, –3)

Plug in values for x and y to see if they work for both equations.

Answer choice (D) is the only answer that works for both equations.

If x = 1
then for y = (−2 × 1) − 1
y = −2 − 1
y = −3

For the second equation:

y = x − 4
−3 = x − 4
−3 + 4 = x − 4 + 4
1 = x

170) The correct answer is: 12

Tip: When you see the sigma sign like this, you need to perform the operation at the right-hand side of the sigma sign. In this problem, we perform the operation for x = 2, x = 3 and x = 4 (because 4 is the number at the top). Then we add these individual products together to get the final result.

For x = 2:   x + 1 = 2 + 1 = 3

For x = 3:   x + 1 = 3 + 1 = 4

For x = 4:   x + 1 = 4 + 1 = 5

3 + 4 + 5 = 12

171) The correct answer is: $\sqrt{61}$

For this type of problem, you will need the distance formula.

$$d = \sqrt{(x_2 - x_1)^2 + (y_2 - y_1)^2}$$

$$d = \sqrt{(6\sqrt{5} - 3\sqrt{5})^2 + (4 - 0)^2}$$

$$d = \sqrt{(3\sqrt{5})^2 + 16}$$

$$d = \sqrt{(9 \times 5) + 16}$$

$$d = \sqrt{45 + 16}$$

$$d = \sqrt{61}$$

172) The correct answer is: $5a^2$

Hopefully you will be comfortable with this type of problem at this point.

Treat the main fraction as division by inverting and multiplying. Then simplify.

$$\frac{a^3/ab}{b/5b^2} = \frac{a^3}{ab} \div \frac{b}{5b^2} = \frac{a^3}{ab} \times \frac{5b^2}{b} = \frac{5a^3b^2}{ab^2} = \frac{ab^2(5a^2)}{ab^2} = 5a^2$$

173) The correct answer is: 20%

For probability problems, your first step is to calculate how many items there are in total, before any are taken away.

Here we have 3 blue scarves, 1 red scarf, 5 green scarves, and 2 orange scarves, so we have 11 scarves in total.

Then deduct the amount that has been taken away. In this problem, one scarf has been removed, so there are 10 scarves remaining.

Since the scarf that was removed was red, there are 2 orange scarves remaining.

So the probability is expressed as a fraction with the remaining pool as the numerator and the remaining total as the denominator, in other words 2/10 in this problem.

Finally we convert this to a percentage: 2/10 = 20%

174) The correct answer is: 36

Look at the relationship between X and Y in order to solve the problem. In each case, we can see that Y = X²

So if X = 6, Y = 36

175) The correct answer is: x^6/y^9

Tip: When raising a power to a power, you have to multiply the exponent outside of the parentheses by the exponents inside the parentheses.

$(x^2 \div y^3)^3 =$

$x^6 \div y^9 =$

x^6/y^9

176) The correct answer is: 5

To solve this problem, you need the following equation:

Triangle area = (base × height) ÷ 2

Now substitute the amounts for base and height.

area = (2 × 5) ÷ 2 =
10 ÷ 2 =
5

177) The correct answer is: 1

Sin^2 is always equal to $1 - cos^2$. In other words, $cos^2 + sin^2 = 1$

178) The correct answer is: 12π

To find the volume of a cone, you need this formula:

Cone volume = (π × radius^2 × height) ÷ 3

Now substitute the values for base and height.

volume = ($\pi 3^2$ × 4) ÷ 3 =
($\pi 9$ × 4) ÷ 3 =
$\pi 36$ ÷ 3 =
12π

179) The correct answer is: 2^6

$2^4 \times 2^2 = 2^{(4+2)} = 2^6$

180) The correct answer is: 74 meters

Set up equations for the areas of the rectangles both before and after the change, using W for the width and L for the length. Then, isolate variable W for the width. Finally, solve by expressing variable W in terms of L.

BEFORE:

2L + 2W = 64

2W = 64 − 2L

W = 32 − L

AFTER:

? = 2(L + 3) + 2(W + 2)

? = 2(L + 3) + 2(32 − L + 2)

? = (2L + 6) + 2(34 − L)

? = 2L + 6 + 68 − 2L

? = 6 + 68

? = 74

181) The correct answer is: $-3x^4 + 7x^3 + 17x^2 - 35x - 10$

Change the positions of the sets of parentheses. Multiply the first term from the first set of parentheses by all of the terms in the second set of parentheses. Then multiply the second term from the first set of parentheses by all of the terms in the second set of parentheses. Then simplify.

$(-3x^2 + 7x + 2)(x^2 - 5) =$

$(x^2 - 5)(-3x^2 + 7x + 2) =$

$(x^2 \times -3x^2) + (x^2 \times 7x) + (x^2 \times 2) + (-5 \times -3x^2) + (-5 \times 7x) + (-5 \times 2) =$

$-3x^4 + 7x^3 + 2x^2 + 15x^2 - 35x - 10 =$

$-3x^4 + 7x^3 + 17x^2 - 35x - 10$

182) The correct answer is: A^{12}

When taking an exponential number to another exponent, you have to multiply the exponents.

$(A^5 \div A^2)^4 =$

$(A^{5-2})^4 =$

$(A^3)^4 =$

A^{12}

183) The correct answer is: 0.25

We have the special operation defined as: $(x \text{ Д } y) = (2x \div 4y)$.

First of all, look at the relationship between the left-hand side and the right-hand side of this equation in order to determine which operations you need to perform on any new equation containing the operation Д and variables x and y.

In other words, in any new equation:

Operation Д is division.
The number or variable immediately after the opening parenthesis is multiplied by 2.
The number or variable immediately before the closing parenthesis is multiplied by 4.

So, the new equation (x Д y) = (2x ÷ 4y) becomes (2 × 8) ÷ (4 × y) = 16

Now solve for (2 × 8) ÷ (4 × y) = 16

(2 × 8) ÷ (4 × y) = 16

16 ÷ 4y = 16

16 = 16 × 4y

16 = 64y

y = 0.25

184) The correct answer is: 24

Permutations are like combinations, except permutations take into account the order of the items in each group. In order to calculate the number of permutations of size S taken from N items, you should use this formula:

N! ÷ (N − S)! =

For the question above:

N = 4 and S = 3

N! ÷ (N − S)! =
(4 × 3 × 2 × 1) ÷ (4 − 3)! =
(4 × 3 × 2 × 1) ÷ 1 =
24 ÷ 1 = 24

185) The correct answer is: $\sqrt{34}$

The length of the hypotenuse is always the square root of the sum of the squares of the other two sides of the triangle.

hypotenuse length C = $\sqrt{A^2 + B^2}$

Now put in the values for the above problem.

$C = \sqrt{A^2 + B^2}$
$C = \sqrt{5^2 + 3^2}$
$C = \sqrt{25 + 9}$
$C = \sqrt{34}$

186) The correct answer is: π/2

To solve this problem, you need these three principles:
(1) Arc length is the distance on the outside (or circumference) of a circle.
(2) The circumference of a circle is always π times the diameter.
(3) There are 360 degrees in a circle.

The angle in this problem is 90 degrees.

360 ÷ 90 = 4; In other words, we are dealing with the circumference of 1/4 of the circle.
Given that the circumference of this circle is 2π, and we are dealing only with 1/4 of the circle, then the arc length for this angle is:

2π ÷ 4 = π/2

187) The correct answer is: 44

Remember that the perimeter is the measurement along the outside edges of the rectangle or other area. If the room is 12 feet by ten feet, we need 12 feet × 2 to finish the long sides of the room and 10 feet × 2 to finish the shorter sides of the room.

(12 × 2) + (10 × 2) = 44

188) The correct answer is: 10 feet by 4 feet

First, we have to calculate the total square footage available. If there are 4 rooms which are 10 by 10 each, we have this equation:

4 × (10 × 10) = 400 square feet in total
Now calculate the square footage of the new rooms.

20 × 10 = 200
2 rooms × (10 × 8) = 160
200 + 160 = 360 total square feet for the new rooms

So the remaining square footage for the bathroom is calculated by taking the total minus the square footage of the new rooms. 400 − 360 = 40 square feet

Since each existing room is 10 feet long, we know that the new bathroom also needs to be 10 feet long in order to fit in. So the new bathroom is 10 feet by 4 feet.

189) The correct answers are: ∠a, ∠d, and ∠f are equal and ∠b, ∠c, and ∠e are also equal.

In problems like this, remember that parallel angles will be equal. So, for example, angles a and d are equal, and angles b and e are equal. Also remember that adjacent angles will be equal when bisected by two parallel lines, as with lines x and y in this problem.

Angles b and c are adjacent, and angles d and f are also adjacent. So, ∠a, ∠d, and ∠f are equal and ∠b, ∠c, and ∠e are also equal.

190) The correct answer is: 16

Circumference = π × radius × 2

The angle given in the problem is 45°. If we divide the total 360° in the circle by the 45° angle, we have: 360 ÷ 45 = 8

So, there are 8 such arcs along this circle. We then have to multiply the number of arcs by the length of each arc to get the circumference of the circle: 8 × 4π = 32π (circumference)

Then, use the formula for the circumference of the circle.

32π = π × 2 × radius
32π ÷ 2 = π × 2 × radius ÷ 2
16π = π × radius
16 = radius

191) The correct answer is: 36

First, calculate the area of the central rectangle. Remember that the area of a rectangle is length times height: 8 × 3 = 24

Using the Pythagorean theorem, we know that the base of each triangle is 4:

$5^2 = 3^2 + base^2$
$25 = 9 + base^2$
$25 - 9 = 9 - 9 + base^2$
$16 = base^2$
$4 = base$

Then calculate the area of each of the triangles on each side of the central rectangle. Remember that the area of a triangle is base times height divided by 2: (4 × 3) ÷ 2 = 6

So the total area is the area of the main rectangle plus the area of each of the two triangles:

24 + 6 + 6 = 36

192) The correct answer is: 30

Remember that the area of a triangle is base times height divided by 2. First, calculate the area of triangle NKM: 6 × (8 + 10) ÷ 2 = 54

Then, calculate the area of the area of triangle NKL: (6 × 8) ÷ 2 = 24

The remaining triangle NLM is then calculated by subtracting the area of triangle NKL from triangle NKM: 54 − 24 = 30

193) The correct answer is: 105°

We know that any straight line is 180°. We also know that the sum of the degrees of the three angles of any triangle is also 180°. The sum of angles X, Y, and Z = 180, so the sum of angle X and angle Z equals 180° − 30° = 150°.

Because the triangle is isosceles, angle X and angle Z are equivalent, so we can divide the remaining degrees by 2 as follows: 150° ÷ 2 = 75°. In other words, angle X and angle Z are each 75°.

Then we need to subtract the degree of the angle ∠XYZ from 180° to get the measurement of ∠WXY: 180° − 75° = 105°

194) The correct answer is: Sin A^2

For any given angle A, Sin A^2 is always equal to 1 − cos A^2 and cos A^2 + sin A^2 = 1

195) The correct answer is: x/y

Here are important trigonometric formulas for calculating the sine, cosine, and tangent of any given angle A:

sin A = x/z
cos A = y/z
tan A = x/y

196) The correct answer is: 4/3

Using the Pythagorean theorem, we know that:

$AB^2 + BC^2 = AC^2$
$AB^2 + 4^2 = 5^2$
$AB^2 + 16 = 25$
$AB^2 + 16 − 16 = 25 − 16$
$AB^2 = 9$
AB = 3

In this problem, the tangent of angle A is calculated by dividing BC by AB.

So the correct answer is 4 ÷ 3 = 4/3

197) The correct answer is: 6.43

The sin of angle Z is calculated by dividing XY by XZ.

sin z = XY/XZ
sin z = XY/10

Since angle Z is 40 degrees, we can substitute values as follows:

sin z = XY/10
0.643 = XY/10
0.643 × 10 = XY/10 × 10
0.643 × 10 = XY
6.43 = XY

198) The correct answer is: θ

If the radius is 1, the radians will be equal to the arc length. So the correct answer is θ.

199) The correct answer is: π ÷ 2 × radians = 90°

The radian is the angle subtended at the center of a circle by an arc that is equal in length to the radius of the circle.

Therefore, the radian is equal to 180 ÷ π , which is approximately 57.2958 degrees.

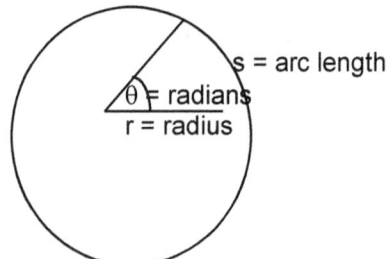

The figure above illustrates the calculation of radians. Remember this formula: θ = s ÷ r

θ = the radians of the subtended angle
s = arc length
r = radius

Also remember these useful equations:

π ÷ 6 × radians = 30°
π ÷ 4 × radians = 45°
π ÷ 2 × radians = 90°

π × radians = 180°
π × 2 × radians = 360°
Since this problem contains a 90 degree angle, the answer is the above equation for 90 degrees: π ÷ 2 × radians = 90°

200) The correct answer is: HF = 2.5 ÷ tan 65°

Since the three locations form a triangle, the length from the hospital to the fire station is calculated by taking the tangent of the angle commencing at the hospital, in this case the tangent of 65°.

tan 65° = FP ÷ HF
tan 65° ÷ tan 65° = 2.5 ÷ HF ÷ tan 65°
1 × HF = 2.5 ÷ (HF × HF) ÷ tan 65°
HF = 2.5 ÷ tan 65°

ADVANTAGE+ EDITION – BONUS MATERIAL

150 ADDITIONAL MATH EXERCISES WITH TIPS AND FORMULAS

Instructions: Complete the math questions that follow by selecting the correct answer from the options provided. You should read the tip following each question before choosing your answer. The answers and solutions are provided after the last exercise. Try to solve as many problems as you can without a calculator; however, at a minimum, you should not use a calculator on problems 1 to 40. You may wish to view the "Math Formula Sheet" at the end of the book for some of the questions.

Arithmetic

1) A company sells electronics online. The annual sales for the first three years of business were: $25,135, $32,787, and $47,004. What were the total sales for the past three years?
   A) $101,326   B) $104,916   C) $104,926   D) $104,944

> This is a question on adding whole numbers. The problem is asking for the total for all three years, so add the three figures together.

2) A customer gives the cashier $50 to pay for the items he purchased, which total $41.28. How much change should be given to the customer?
   A) $7.82   B) $8.18   C) $8.27   D) $8.72

> This is a question on subtracting whole numbers. To calculate the change, you need to take the amount of money the customer gives the cashier and subtract the amount of the purchase.

3) A car salesperson earns a $175 referral fee on every customer who accepts a customer service upgrade. The salesperson referred 8 customers for the service upgrade this month. How much did the salesperson earn in referral fees for the month?
   A) $1050   B) $1200   C) $1225   D) $1400

> This is a question on multiplying whole numbers. Multiplication problems will often include the words 'each' or 'every.' Multiply the amount of the referral fee by the number of customers to solve.

4) An employee's weekly pay is $535.50 and she works 30 hours per week. How much is she paid per hour?
   A) $17.83   B) $17.84   C) $17.85   D) $18.34

> This is a question on dividing whole numbers. Division problems will often include the word 'per.' Divide the total weekly amount by the number of hours to solve.

5) Business losses are represented as negative numbers, while business profits are represented as positive numbers. A business makes the following profits and losses during a four week period: –$286, $953, $1502, and –$107. What was the total business profit or loss during these four weeks?
   A) $2,026   B) $2,062   C) $2,080   D) –$2,026

> This is a question on adding negative numbers. When you have to add a negative number to a positive number, you are subtracting. So, add the business profits and subtract the business losses to solve.

6) Location below sea level is represented as a negative number. The location below sea level of Lake Alto is –35 meters. The location below sea level of Lake Bajo is 62 meters deeper than Lake Alto. What figure represents the location below sea level for Lake Bajo?
   A) –97   B) 97   C) –62   D) –27

> This is a question on subtracting negative numbers. The facts state that the location below sea level of Lake Bajo is 62 meters deeper than Lake Alto, so we need to subtract this figure from the location below sea level of Lake Alto. The location below sea level of Lake Alto is a negative number, so you are subtracting a negative from a negative.

7) A company has completed 3/5 of a project. What figure below expresses the project completion amount as a decimal number?
   A) 0.06   B) 0.60   C) 1.67   D) 3.00

> This is a question on changing fractions to decimals. To express a fraction as a decimal, treat the line in the fraction as the division symbol and divide accordingly. Remember to be careful with the decimal placement in your final answer.

8) A teacher reports attendance as a decimal figure, calculated as the number of students attending divided into the total number of students in the class. This week, the attendance was 0.55. What percentage best represents the attendance for this week?
   A) 0.55%   B) 5.50%   C) 55.0%   D) 55.5%

> This is a question on changing decimals to percentages. To express a decimal number as a percentage, move the decimal point two places to the right. Then add the percent sign.

9) An employee has used up 5/14 of his vacation days. Approximately what percentage of vacation days has the employee already used?
   A) 0.357%   B) 2.800%   C) 3.571%   D) 35.7%

> This is a question on changing fractions to decimals. Treat the line in the fraction as the division symbol and divide. Then move the decimal point two places to the right, and add the percent sign.

10) A driver has used 0.75 of the gas he last put in his car. What fraction best represents the amount of gas used?
    A) 1/4   B) 2/5   C) 3/5   D) 3/4

> This is a question on changing a decimal number to a fraction. You should be able to recognize the equivalent decimal numbers for commonly-used fractions such as ½ or ¾ for your exam. If you are unsure, perform division on the answer choices to solve.

11) It is reported that 33% of all new stores close within five years of opening. What fraction best represents this percentage?
A) 1/3   B) 1/4   C) 1/5   D) 2/3

This is a question on changing a percentage to a fraction. You should be able to recognize the equivalent fractions for commonly-used percentages for the test. If you are unsure of the answer, perform division on the answer choices to solve.

12) A carpet store is offering 45% off in a sale this month. What decimal number below best represents the percentage off?
A) 0.0045   B) 0.0450   C) 0.4500   D) 4.5000

This is a question on changing percentages to decimals. Any given percentage is out of 100%, so we divide by 100 to express a percentage as a decimal. So, move the decimal point two places to the left and remove the percent sign.

13) A bakery has to pay 36 cents for each pound of flour it buys. It decides to buy 14 1/4 pounds of flour today. How much will it have to pay?
A) $3.60   B) $5.13   C) $5.31   D) $142.50

This is a question on calculations involving units of money. Express both amounts as decimal numbers and multiply to solve.

14) A bookkeeper has just been with a client for 0.35 hours. Approximately how many minutes did the bookkeeper spend with this client?
A) 3.5 minutes   B) 5.8 minutes   C) 21 minutes   D) 35 minutes

This is a question on calculations involving units of time. There are 60 minutes in an hour, so multiply the minutes in the hour by the decimal number given in the problem to solve.

15) A flower store charges $24 for a small arrangement of flowers. A customer will get a $5 discount if he or she provides his or her own vase for the small arrangement. This month, there were 12 customers who ordered small arrangements and provided their own vases. How much money in total did the flower store make on arrangements sold to these 12 customers?
A) $228   B) $282   C) $288   D) $348

This is a question with two operations. Subtract the discount from the original price. Then multiply this figure by the number of units sold.

16) A bricklayer works for a construction company. He laid bricks for 7 hours per day for 4 days on one job. The customer was billed $45 per hour for his work, and he was paid $25 per hour. After the bricklayer's wages have been paid, how much money did the company make for his work on this job?
A) $175   B) $180   C) $315   D) $560

> This question has three operations. First, you need to determine the total number of hours worked for the 4 days. Then calculate the profit the company makes per hour. Finally, multiply the total number of hours worked by the profit per hour to solve.

17) A pharmacist owns a local drug store. Last week, she filled 250 prescriptions in 40 hours. Assuming that each prescription takes the same amount of time, how many minutes should it take her to fill a single prescription?
   A) 9.6 minutes B) 6.25 minutes C) 3.75 minutes D) 0.16 minutes

> This is a question with two operations. Since there are 60 minutes in an hour, we multiply by 60 to get the number of minutes. Then divide by the number of prescriptions to get the rate.

18) A truck driver delivered 120 orders this week. She delivered 105 of the orders on time. What percentage of the driver's orders was delivered on time?
   A) 0.875% B) 8.75% C) 87.5% D) 0.125%

> This is a question with two operations. Take the amount of orders that were delivered on time and divide by the amount of total orders. Then convert to a percentage.

19) A scientist measures cell growth or attrition. Each day a measurement is taken. Cell growth is represented as a positive figure, while cell attrition is represented as a negative figure. On Monday cell growth was 27, and for all days Tuesday through Friday, cell attrition was 13 per day. What number represents total cell growth or attrition for these five days?
   A) 25 B) –25 C) 40 D) –40

> This is a question on multiplying negative numbers. Cell attrition is a negative number, so perform multiplication to get the total for Tuesday through Friday. Then add the cell growth for Monday to solve.

20) A vegetable farmer works until noon each day. The chart below shows the amounts of cucumbers per hour that she picked one morning:
   7:00 to 8:00   23 cucumbers   10:00 to 11:00 24 cucumbers
   8:00 to 9:00   25 cucumbers   11:00 to 12:00 22 cucumbers
   9:00 to 10:00   26 cucumbers

   On average, how many cucumbers did the farmer pick per hour?
   A) 23 B) 24 C) 25 D) 26

This is a question on calculating averages. The average is sometimes called the arithmetic mean, so you may see both terms on the test. To find the average, you need to add all of the amounts to get the total, and then divide the total by the number of hours.

21) A local company makes one particular kind of concrete. For this concrete, 2 units of sand have to be added to every 3 units of cement powder used. A batch of this concrete that has 66 units of cement powder is being made. How many units of sand should be added to this batch?
A) 2 B) 3 C) 22 D) 44

This is a question on ratios. Take the 66 units of cement powder for the current batch and divide by the 3 units stated in the original ratio. Then multiply this result by the 2 units of sand stated in the original ratio to solve.

22) It is company policy that the ratio of employees to supervisors should be 50:1. So, for every 50 employees in a company, there should be 1 supervisor. If there are 255 employees, how many supervisors are there?
A) 1 B) 2 C) 3 D) 5

This is another question on ratios. The problem states that we are working with a ratio, so the employees and the supervisors form separate groups. First, add the two groups together. Then take the total number of employees stated in the problem and divide this by the figure you have just calculated to get the number of supervisors.

23) A report shows that 2 out of every 20 employees in a particular company are interested in applying for a promotion. If the company has 480 employees in total, how many employees are interested in applying for a promotion?
A) 20 B) 24 C) 42 D) 48

This is a question on proportions. Problems on proportions often use the phrase 'out of.' The problem uses the phrase '2 out of every 20 employees' so we know that there are 2 employees who form a subset within each group of 20. So, take the total number of employees and divide this by 20. Then multiply this result by the amount in the subset to solve.

24) A mechanic spent from 8:10 to 8:22 changing three wheel covers on a car. At this rate, how many wheel covers could he change per hour?
A) 3 B) 5 C) 15 D) 20

This is a question on calculating rate. Calculate the amount of time in minutes that was spent on the three wheel covers. Then calculate the time in minutes needed to change 1 wheel cover. Then divide this amount into 60 minutes to solve.

© COPYRIGHT 2007, 2011. Academic Success Media © COPYRIGHT 2020. Academic Success Group.
This material may not be copied or reproduced in any form.

25) A fencing company put up 15²/₈ yards of fence for one customer and 13⁵/₈ yards of fence for another customer. How many yards of fence did the company put up for both customers in total?
   A) 28³/₈   B) 28⁵/₈   C) 28⁷/₈   D) 28⁷/₁₆

> This is a question on adding fractions that have a common denominator. First, add the whole numbers that are in front of each fraction. Then add the fractions. If you have two fractions that have the same denominator, which is the number on the bottom of the fraction, you add the numerators and keep the common denominator. Then combine the new whole number and the new fraction to solve.

26) A food company fills gourmet food boxes with various products. So far today, 2³/₈ boxes have been filled for one order and 4¹/₈ boxes have been filled for another order. How many total boxes have been filled so far today?
   A) 6¹/₂   B) 6¹/₄   C) 6³/₄   D) 6³/₁₆

> This is another question on adding fractions that have a common denominator. Follow the same steps as for the previous question, but also simplify the fraction to solve. This means that you have to reduce the numerator and denominator by dividing them by the same number, which is known as a common factor.

27) A customer has just placed an order to have an awning made for his front window. According to the measurements, 5³/₁₆ yards of canvas will be needed to make the awning. However, the customer calls later to say that his initial measurement was incorrect, and only 4¹/₁₆ yards of canvas is actually needed to make the awning. Which fraction below represents the amount by which the amount of canvas has been reduced?
   A) 1¹/₈   B) 1¹/₁₆   C) 1¹/₃₂   D) 1³/₁₆

> This is a question on subtracting fractions with a common denominator. First, subtract the whole numbers, and then subtract the fractions. If you have two fractions that have the same denominator, you subtract the numerators and keep the common denominator. Then simplify the fraction. Finally, combine the whole number and the simplified fraction to solve.

28) Certain additives need to be placed in a bottle to make a particular product. The company measures each additive in decimal units, with 1 unit representing the filled bottle. The bottle contains 0.25 units of additive A, 0.50 units of additive B, and 0.10 units of additive C. Which of the following represents, in terms of units, how full the bottle currently is?
   A) 08.5   B) 0.85   C) 0.90   D) 0.95

> This is a question on adding commonly-known decimals. Add the three figures together to solve. Remember to be sure to put the decimal point in the correct place when you work out the solution.

29) A recent survey shows that 50% of your customers rated your service as excellent and 25% rated your service as very good. What percentage below represents the total amount of customers who rated your service either excellent or very good?
   A) 25%   B) 50%   C) 75%   D) 85%

This is a question on adding commonly-known percentages. Simply add the percentages together to solve.

30) A customer has just ordered 5 units of a product. Each unit of the product takes 1¼ hours to make. How much time is needed to make this order?
A) 5 hours and 25 minutes
B) 5 hours and 55 minutes
C) 6 hours and 4 minutes
D) 6 hours and 15 minutes

This is a question on multiplying a mixed number by a whole number of units. First, multiply the whole numbers. Then multiply the whole number of units by the fraction. Then convert this improper fraction to a mixed number. Add the whole number and the mixed number, and convert to hours and minutes to solve.

31) A dressmaker who works in a tailoring shop is trying to decide what setting to use on the sewing machine. She has tried the 1/8 inch stitch but has realized that it is too small. The stitches on the machine are sized in 1/32 increments. What size stitch should she try next?
A) 3/16
B) 5/32
C) 6/16
D) 6/32

This is a question on performing calculations on fractions with different denominators. Convert 1/8 to the following equivalent fraction: 1/8 = ?/32

32) Amal runs a souvenir store that sells key rings. She can get 50 key rings from her first supplier for 50 cents each. She can get the same 50 keys rings from her second supplier for $30 in total or from her third supplier for $27.50. How much will she pay if she gets the best deal?
A) $25.00
B) $25.25
C) $25.50
D) $27.50

This is a question on finding the best deal when you have to perform a one-step calculation. Read the facts carefully, work out the total prices for all three suppliers, and then compare prices.

33) A budget hotel charges $45 per night and $280 per week. If a guest stays at the hotel for 9 nights, what is the least that he will pay for his stay?
A) $280
B) $315
C) $325
D) $370

This is a question on finding the best deal when you have to perform two-step calculations. Determine the duration of the stay in weeks and nights. Then add the cost for 1 week to the cost for 2 days to solve.

34) The price of an item is normally $15, but customers with a loyalty card can purchase it at the discounted price of $12. What percentage best represents the discount awarded to these customers?
A) 3%
B) 5%
C) 15%
D) 20%

> This is a question on calculating the percentage of a discount. Divide the dollar amount of the discount by the original price to get the percentage of the discount.

35) A retail ceramics store sells mugs and bowls. It buys one type of mug for $3 and sells it for $9. It uses the same percentage mark up on one type of bowl that it buys for $4. What figure below represents the sales price of the bowl?
A) $6
B) $8
C) $12
D) $16

> This is a question on calculating a mark-up. You need to calculate the percentage for the mark-up on the first product and apply this percentage mark-up to the second product. Remember to use the percentage mark-up, rather than a dollar value. You may need the following formulas if you don't already know how to calculate mark-up: Dollar value of mark-up = Sales price in dollars − Cost in dollars; Percentage mark-up = Dollar value of mark-up ÷ Cost in dollars

36) A company got $20 off of an order. This amounted to a 25% discount off the order. What would the company have paid without the discount?
A) $4
B) $5
C) $25
D) $80

> This is a question on calculating a reverse percentage. To calculate a reverse percentage you need to divide, rather than multiply. So, take the dollar value of the discount and divide by the percentage to solve.

37) A company that fabricates cleaning products begins to make the first batch of products on Monday at 10:30 am. The actual production time is 3 hours and 25 minutes. This is followed by a bottling and labeling process that takes 1 hour and 40 minutes and a packaging process that takes a further 26 hours. If production keeps to this schedule, when will the first batch be ready for shipment?
A) Tuesday at 12:30 pm
B) Tuesday at 3:55 pm
C) Tuesday at 5:35 pm
D) Wednesday at 3:55 pm

> This is a question on calculating the hours and minutes that have passed since the start of a job or process. Calculate the total time for the entire process and add to the starting time to solve.

38) Maria sells soft drinks in a convenience store that she runs. She can buy 240 soft drinks from one supplier for 25 cents each or from a different supplier for $58 for all 240 drinks. Both suppliers are in the same state, so she has to pay a sales tax of 6.5% on either purchase. If she chooses the best price for the soft drinks, including tax, how much will she pay in total?
A) $58.00
B) $60.00
C) $61.77
D) $63.90

> This is another question on finding the best deal. Remember to add the dollar amount of the sales tax to both calculations for this problem.

39) A picture framing store can make 20 small frames in 4 days or 21 large frames in 3 days. A customer has just placed an order with for 40 small frames and 64 large ones. Approximately how many days will it take to make them all?
   A) 7 B) 11 C) 14 D) 17

This is a question on calculating production rates by unit. Determine the unit rates per day for each of the products by dividing the output by the number of days. Then divide the rates into the amount of items ordered to solve.

40) The report on a production order shows that 12.5% of the work has been completed in the past 4 days. If work continues at the same rate, how many more days will be required in order to finish the order?
   A) 3 B) 4 C) 28 D) 32

This is a question on calculating rate by time. Calculate the percentage of work completed per day, and then determine how many days are needed for the job.

Geometry

41) A land surveyor must measure the distance between landmarks. She has measured a distance between two landmarks and discovered that it is 538 feet. What is the approximate distance between the landmarks in terms of meters?
   A) 45 B) 164 C) 1367 D) 1765

This is a question on using a formula with a measurement. Use the following formula and multiply to solve: 1 foot = 0.3048 meters

42) A physical therapist measures how far her clients are able to walk during each session. One client walked 123 feet and 6 inches during his first session and 138 feet and 8 inches during his second session. What is the combined total of the distance walked for the two sessions?
   A) 261 feet 24 inches   C) 262 feet 8 inches
   B) 261 feet 6 inches   D) 262 feet 2 inches

This is a question on performing a calculation with mixed units. It is usually easiest to perform one calculation with the feet and another with the inches. You may need to convert the total inches back to feet and inches if there are more than 12 inches in the second calculation.

43) A nutritionist advises clients and sells supplements to them. A box containing the supplements weighs 8 pounds and 5 ounces when full. The box itself weighs 7 ounces when it is empty. Each supplement weighs 0.75 ounces. About how many supplements should be in the box?
   A) 168 B) 177 C) 178 D) 186

> This is a question on performing conversions within systems of measurement. Here we have to convert between pounds and ounces. Convert the total weight of the product (excluding the box weight) to ounces then divide the total ounces by the ounces per unit to solve. 1 pound = 16 ounces

44) A garden store fertilizes and treats customers' lawns. One customer wants to fertilize and treat his lawn, which is 50¼ feet by 60¼ feet in size. The cost of the fertilizer and treatment is $5.25 per square yard. To the nearest dollar, how much will it cost the customer to fertilize and treat his lawn?
   A) $177   B) $1,766   C) $5,298   D) $15,895

> This is a question on working with quantities that contain fractions. Convert the mixed numbers to decimals and multiply. Then convert to square yards and solve. 1 square yard = 9 square feet

45) It is company policy to have at least 60 yards of dark black yarn in stock at the start of every month. Inventory has been taken this morning and there are 2 balls of dark black yarn that are 75 inches each and 4 balls of dark black yarn that are 25¼ inches each in stock. This yarn must be purchased in 5-yard-long balls. How many balls of yarn should be purchased in order to replenish the stock?
   A) 10   B) 11   C) 33   D) 36

> This is a question on working with fractional units. Calculate the amount of remaining stock in inches, and then convert from inches to yards. Then calculate the amount required to restock. Remember that it is not possible to buy a fractional part of a ball, so you have to round up to solve.

46) A company that manufactures liquid cosmetics needs to test a 0.75 gram sample of an active ingredient of a liquid cosmetic. The correct concentration ratio is 50 milligrams of active ingredient to 1.5 milliliters of liquid. How many milliliters of liquid should be added to the sample?
   A) 0.000015   B) 0.000225   C) 15.0   D) 22.5

> This is a question on converting grams to milligrams. Convert to grams (1 gram = 1,000 milligrams). Then apply the correct ratio to solve.

47) Find the midpoint of the line segment that connects the points (5, 2) and (7, 4).
   A) (6, 3)   B) (3, 6)   C) (3.5, 5.5)   D) (12, 6)

48) If store A is represented by the coordinates (−4, 2) and store B is represented by the coordinates (8, −6), and store A and store B are connected by a line segment, what is the midpoint of this line?
   A) (2, 2)   B) (2, −2)   C) (−2, 2)   D) (−2, −2)

> The midpoint of two points on a two-dimensional graph is calculated by using the midpoint formula:
>
> $$(x_1 + x_2) \div 2 , (y_1 + y_2) \div 2$$

49) What is the distance between (2, 3) and (6, 7)?
   A) 4
   B) 16
   C) $\sqrt{16}$
   D) $\sqrt{32}$

> The distance formula is used to calculate the linear distance between two points on a two-dimensional graph. The two points are represented by the coordinates (x_1, y_1) and (x_2, y_2).
>
> $$d = \sqrt{(x_2 - x_1)^2 + (y_2 - y_1)^2}$$

50) The measurements of a mountain can be placed on a two-dimensional linear graph on which x = 5 and y = 315. If the line crosses the y axis at 15, what is the slope of this mountain?
   A) 60
   B) 63
   C) 300
   D) 315

> The slope formula: $m = \dfrac{y_2 - y_1}{x_2 - x_1}$
>
> The slope-intercept formula: y = mx + b, where m is the slope and b is the y intercept.
>
> Now use these formulas to solve the graph problem that follows.

51) Which of the following statements is true with respect to the lined graph below?

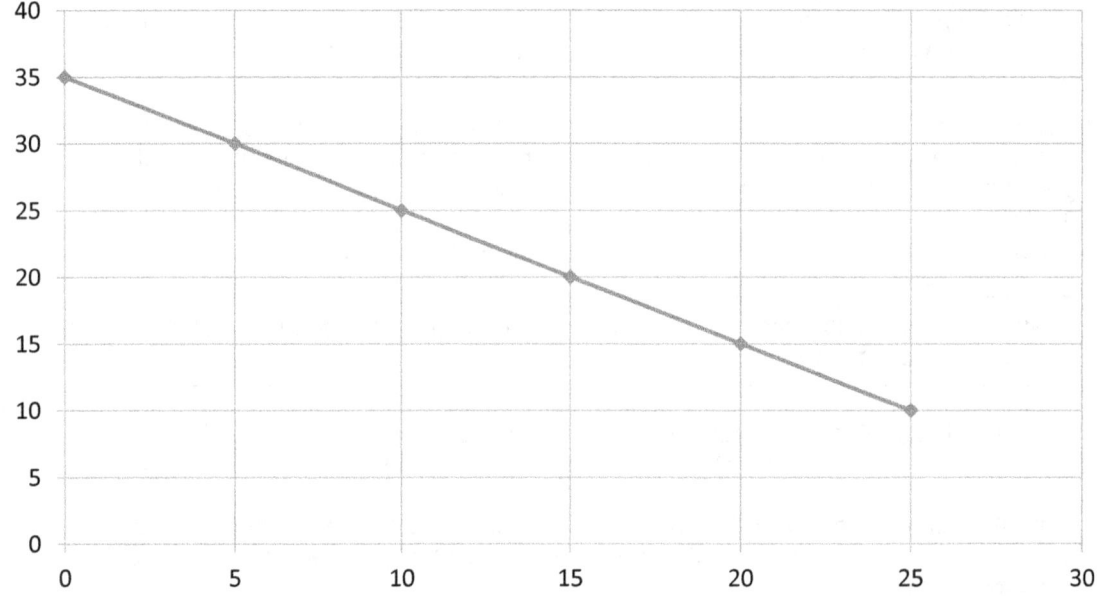

A) The line has a slope of −1 and contains point (20, 15).
B) The line has a slope of 1 and contains point (20, 15).
C) The line has a slope of −1 and contains point (15, 20).
D) The line has a slope of 1 and contains point (15, 20).

52) State the x and y intercepts that fall on the straight line represented by the equation:
y = x + 14
A) (−14, 0) and (0, 14)
B) (0, 14) and (0, −14)
C) (14, 0) and (0, −14)
D) (0, −14) and (14, 0)

53) Find the x and y intercepts of the following equation: $x^2 + 2y^2 = 144$
A) (12, 0) and (0, $\sqrt{72}$)
B) (0, 12) and ($\sqrt{72}$, 0)
C) (0, $\sqrt{72}$) and (0, 12)
D) (12, 0) and ($\sqrt{72}$, 0)

For questions like the two previous ones on x and y intercepts, substitute 0 for y in the equation provided to find the value of x. Then substitute 0 for x to find the value of y and solve the problem.

54) A carpenter creates triangular-shaped corner shelves from oak and other wood for sale to furniture and home stores. He needs to report the area of each shelf to the buyer as part of the sales agreement. He needs to calculate the area of a triangular-shaped shelf that has a base of 12 inches and a height of 14 inches. What is the area of this shelf in square inches?
A) 56
B) 84
C) 168
D) 1728

Use the formula for the area of a triangle: ½ (base × height)

55) Triangle ABC is a right-angled triangle. Side A and side B form the right angle, and side C is the hypotenuse. If A = 3 and B = 2, what is the length of side C?
A) 5
B) $\sqrt{5}$
C) $\sqrt{13}$
D) 13

The hypotenuse is the side of the triangle that is opposite the right angle. To calculate the length of the hypotenuse in right triangles, you will need the Pythagorean Theorem. According to the theorem, the length of the hypotenuse (represented by side C) is equal to the square root of the sum of the squares of the other two sides of the triangle (represented by A and B). For any right triangle with sides A, B, and C, you need to remember this formula:

$$\text{hypotenuse length } C = \sqrt{A^2 + B^2}$$

56) A carpenter is making a special triangular-shaped corner shelf for a custom order. The customer lives in a 300-year-old house, so the walls are not completely straight and the corners are not completely square. He needs to make a triangular shelf that will have one 44° angle and one 47° angle. What is the measurement in degrees of the third angle of this shelf?
A) 45°
B) 45.5°
C) 89°
D) 90°

The sum of all three angles in any triangle is always equal to 180 degrees.

57) A real-estate developer has recently purchased a circular-shaped tower. The first floor of the building has been divided into 5 pie-shaped segments that join at the center of the circle. The first segment measures 82° along the outside edge. The second segment has a measurement of 79°, the third has a measurement of 46° and the fourth has a measurement of 85°. What is the measurement in degrees of outside edge the fifth segment?
A) 48   B) 49   C) 58   D) 68

A complete circle measures 360 degrees.

58) A building project has a circular tower. The floor of the tower, which has a 12-foot radius, needs to be filled in with concrete. In order to do this, the area of the floor of the tower needs to be calculated. What is the approximate area of the floor of the tower in square feet?
A) 452.16   B) 376.80   C) 226.08   D) 37.68

This is a question on calculating the area of a circle. The formula for the area of a circle is as follows: circle area = $\pi \times (radius)^2 \approx 3.14 \times (radius)^2$

For all of the problems that follow, use 3.14 for π unless otherwise indicated.

59) A technician measures the wear on tractor tires. In order to determine the rate of wear, the circumference of each tire must be determined first. The tire currently being measured has a diameter of 46.5 inches. What is the circumference?
A) 23.500 inches   B) 73.005 inches   C) 146.01 inches   D) 292.02 inches

This is a question on calculating the circumference of a circle. Circumference = $\pi \times$ diameter $\approx 3.14 \times$ diameter

60) Becky is making a patchwork quilt that is going to be 6 feet long and 5 feet wide. What will the surface area of the quilt be in square feet?
A) 11   B) 22   C) 25   D) 30

This is a question on calculating the area of a rectangle. Area of a rectangle = length × width

61) A fence needs to be put around a field that is 12 yards long and 9 yards wide. What figure below best represents the perimeter of this field in yards?
A) 21   B) 42   C) 54   D) 72

This is a question on calculating the perimeter of a rectangle. Remember not to confuse area and perimeter as they are different calculations. Perimeter of a rectangle = 2 × (length + width)

62) A circular fish pond is being designing for your local park. The pond has an area of about 78.5 square feet. What is the approximate diameter of the pond?
   A) 5 feet	B) 10 feet	C) 15.7 feet	D) 25 feet

> You need to use the formula in reverse for this question, so use 3.14 for π and divide by 3.14, instead of multiplying by 3.14. Remember that diameter is double the radius, so if the radius is 3 feet, for example, the diameter is 6 feet. Remember that the formula is: circle area ≈ 3.14 × (radius)^2.

63) A rectangular vegetable garden has an area of 360 square feet. If the length of the garden is 30 feet, what is the width of the garden?
   A) 12 feet	B) 24 feet	C) 115 feet	D) 150 feet

> This is another question on rearranging a formula. The area of a rectangle = length × width. Here, we are given the area, so we need to divide that by the length to solve.

64) A tank that holds dye is 5 feet wide, 8 feet long, and 3 feet high. How many cubic feet of dye can the tank hold when it is completely full?
   A) 15	B) 24	C) 40	D) 120

> The length, width, and height are different measurements, so we need the formula for the volume of a rectangular solid or box: volume = *length* × *width* × *height*

65) A cube footrest has a side length of 18 inches. How many cubic inches of filling should be placed inside the footrest?
   A) 5,832	B) 729	C) 324	D) 72

> For this problem, we need to calculate the volume of a cube. The formula for the volume of a cube is as follows: cube volume = (*length of side*)^3

66) A company processes dairy products. Milk is stored in a spherical storage tank that is 72 inches across on the inside. The tank is now 80% full of milk. What is the volume of the milk in the tank in cubic inches?
   A) 156,267	B) 156,627	C) 159,333	D) 195,333

> This is a question on calculating volume. You need the following formula: Volume of a sphere = 4/3 × π × radius^3 ≈ 4/3 × 3.14 × radius^3. Use the formula and multiply by the percentage stated in the problem.

67) A cylindrical tank has a 5 meter radius and is 21 meters in height. What is the volume of the tank?
   A) 329.70	B) 1648.5	C) 549.50	D) 659.40

> This is another question on calculating volume. Cylinder volume = π × radius^2 × height ≈ 3.14 × radius^2 × height. Substitute the values into the formula, and then perform the operations in the formula to solve.

68) A confection company manufactures three different sizes of ice cream cones. The large cones are 6 inches high and have a 1.5 inch radius, the medium cones are 5 inches high and have a 1 inch radius, and the small cones are 4 inches high and have a 0.5 inch radius. What is the difference between the volume in cubic inches of the large cone and the medium cone?
A) 4.19   B) 5.23   C) 8.90   D) 14.13

This is a question on calculating differences in volumes. Cone volume = (π × radius² × height) ÷ 3 ≈ (3.14 × radius² × height) ÷ 3. Calculate the difference between the volumes of the two cones to solve.

69) A building contractor is laying wooden parquet pieces on a floor. The wooden part of the floor will cover an area that measures 8 feet long by 4 feet wide. Each wooden parquet piece measures 12 inches by 6 inches. What is the minimum number of wooden parquet pieces that will be needed in order to cover the wooden part of the floor?
A) 16   B) 32   C) 48   D) 64

Determine how many wooden pieces will fit along the length of the floor. Next, determine how many wooden pieces can fit along the width. Finally, multiply to solve.

70) A painter is painting a wall that is 16 feet long and 11 feet high. She needs to calculate the surface area of the wall in order to know how much paint to buy. What is the surface area of the wall in square feet?
A) 54   B) 121   C) 176   D) 256

Don't let the fact that this is a wall confuse you. You still need to calculate the area. Which previous formula should you use?

71) A rectangular solid container needs to be filled with a liquid substance. The length of the rectangular solid is 12 feet, the width is 9 feet, and the volume is 1080 cubic feet. What is the height of the rectangular solid?
A) 10 feet   B) 12 feet   C) 90 feet   D) 100 feet

To calculate the volume of a rectangular sold or box, use this formula: *length × width × height*. You are doing the formula in reverse, so you need to divide by 12 and then divide that result by 9 to solve.

72) A beaker is cylindrical and measures 18 inches high and 12 inches in diameter. However, the volume has to be converted from cubic inches to gallons for a report. What is the approximate volume of the beaker in terms of gallons?
A) 2.9 gallons   B) 8.8 gallons   C) 10.4 gallons   D) 8,138.88 gallons

The formula for the volume of a cylinder is: volume = π × (*radius*)² × *height* ≈ 3.14 × (*radius*)² × *height*. To convert cubic inches to gallons: 1 gallon = 231 cubic inches

73) The volume of a cube-shaped object needs to be calculated. The cube has a side length of 9 feet. However, a report is asking for the volume of the object in terms of cubic inches. Which figure below should be used?
A) 729 cubic inches
B) 1,728 cubic inches
C) 139,968 cubic inches
D) 1,259,712 cubic inches

The volume of a cube = (length of side)^3. 1 cubic foot = 1.728 cubic inches

74) A company ships products overseas in large rectangular shipping containers. One type of container is 25 feet long, 12 feet wide, and 18 feet high. The container is currently 75% full of a particular product. What is the volume in cubic yards of the product in the container?
A) 150 cubic yards   B) 200 cubic yards   C) 405 cubic yards   D) 4,050 cubic yards

Use the formula for calculating the volume of a rectangular solid. Your result will be in cubic feet. Then convert to cubic yards to solve. (1 cubic yard = 27 cubic feet)

75) A company manufactures glue and other adhesives that contain a chemical called PVA. At least 50 quarts of PVA need to be in stock at the start of every month. Inventory has been taken this morning and there are 2 containers of PVA that hold 16 cups and 7 ounces each. There are also 3 containers of PVA that hold 20 cups and 4 ounces each. This PVA must be purchased in 5-quart containers. How many containers are needed in order to replenish the stock?
A) 0   B) 5   C) 6   D) 7

If you do not know the relationships between ounces and cups, and between cups and quarts, please look at the formula sheet in the appendix.

76) A company that manufactures hand soap and laundry detergent has to order liquid parabens that are used in its products. The parabens are stored in two identically sized vats. The vats measure 10 feet by 10 feet by 12 feet. The first vat is 3/4 full and the second vat is 4/5 full. The parabens cost 12 cents a cubic inch. To the nearest dollar, what is the value of the parabens in the two vats?
A) $223   B) $3,857   C) $4,977   D) $385,690

Use the formula for calculating the volume of a rectangular solid. Your result will be in cubic feet. Then convert to cubic inches to solve. (1 cubic foot = 1,728 cubic inches)

77) A company that manufactures batteries stores acid in a conical-shaped container that is 6 feet in diameter and 8 feet in height. The manager has calculated that the inside of the container at its maximum could contain approximately 226 cubic feet of acid. What error, if any, has been made in this calculation?
A) There is no error in the calculation.
B) The manager forgot to divide by 3.
C) The manager forgot to multiply by 3.14.
D) The manager squared the container's diameter instead of its radius.

Perform the calculations in answers B, C, and D to isolate the error and solve.

78) An electrician installs wiring and lighting in new homes. The client would like to install lights on the walls in the living room. The living room is 25 feet long and 10 feet wide. The client would like a light to be installed on each wall in 5-foot increments. However, no lights are to be installed in the corners of the room. How many lights will be needed in order to carry out this job?
A) 8   B) 10   C) 12   D) 14

This is a question on increments in perimeter. You should draw a diagram on scratch paper to help you answer.

79) A company that manufactures ice cubes and frozen refreshments makes two sizes of ice cubes. The large ice cubes have a side length of 1.8 millimeters, and the small ice cubes have a side length of 1.4 millimeters. What is the amount in cubic millimeters of the difference in volume between the large ice cube and the small one?
A) 0.064   B) 1.960   C) 2.744   D) 3.088

Calculate the volume of each cube. Then subtract these two results to solve. Remember that the volume of a cube = (length of side)^3.

80) A building engineer has been asked to calculate the areas of two triangular shapes. The large triangle has a base of 12 inches and a height of 18 inches. The small triangle has a base of 8 inches and a height of 14 inches. What is the difference in the areas of the two shapes?
A) 8   B) 16   C) 25   D) 52

Use the formula for the area of a triangle that we have seen in a previous question in this section. Then subtract to solve.

Quantitative Reasoning and Statistics

81) A student receives the following scores on his exams during the semester: 89, 65, 75, 68, 82, 74, 86. What is the mean of his scores?
A) 24   B) 74   C) 75   D) 77

The arithmetic mean is the same thing as the arithmetic average. In order to calculate the mean, add up the values of all of the items in the set, and then divide by the number of items in the set.

82) Members of a weight loss group report their individual weight loss to the group leader every week. During the week, the following amounts in pounds were reported: 1, 1, 3, 2, 4, 3, 1, 2, and 1. What is the mode of the weight loss for the group?
A) 1 pound   B) 2 pounds   C) 3 pounds   D) 4 pounds

This is a question on mode. Mode is the value that occurs most frequently in a data set. For example, if 10 students scored 85 on a test, 6 students scored 90, and 4 students scored 80, the mode score is 85.

83) Mark's record of times for the 400-meter freestyle at swim meets this season is: 8.19, 7.59, 8.25, 7.35, and 9.10. What is the median of his times?
A) 7.59   B) 8.19   C) 8.25   D) 8.096

84) Find the median of the following data set: 10, 12, 8, 2, 5, 21, 8, 6, 2, 3
    A) 7   B) 6.5   C) 2   D) 19

These two questions are asking you to find the median of a set of numbers. If there is an odd number of items in the data set, the median is the number that is in the middle of the set when the numbers are in ascending order. If there is an even number of items in the data set, we have to take the average of the two numbers that are in the middle of the set when the numbers have been placed in ascending order.

85) A student receives the following scores on her assignments during the term: 98.5, 85.5, 80.0, 97, 93, 92.5, 93, 87, 88, 82. What is the range of her scores?
    A) 17.0   B) 18.0   C) 18.5   D) 89.65

This is a question on calculating range. To calculate range, the lowest value in the data set is deducted from the highest value in the data set.

86) What is the mode of the numbers in the following list? 1.6, 2.9, 4.5, 2.5, 5.1, 5.4
    A) 3.5   B) 3.1   C) 3.0   D) no mode

This is another question on mode. What happens to the mode if no number in the set occurs more than once?

87) There are 10 cars in a parking lot. Nine of the cars are 2, 3, 4, 5, 6, 7, 9, 10, and 12 years old, respectively. If the average age of the 10 cars is 6 years old, how old is the 10th car?
    A) 1 year old   B) 2 years old   C) 3 years old   D) 4 years old

This is a question on how to calculate the missing value from the calculation of the mean. We don't know the age of the 10th car, so set up an equation and put this in as x to solve:
$(2 + 3 + 4 + 5 + 6 + 7 + 9 + 10 + 12 + x) \div 10 = 6$

88) 100 participants took an intelligence test. The mean score for the first 50 participants was 82, and the mean score for the next 50 participants was 89. What is the mean test score for all 100 participants?
    A) 85.5   B) 86.5   C) 87   D) 88

Find the total points for the first group. Then find the total points for the second group. Add these two results together to find the total points for all the participants. Then divide the total points by the total number of members in the group.

89) An employee at the Department of Motor Vehicles wanted to find the mean of the ten driving theory tests he administered this morning. However, the employee divided the total points from the ten tests by 8, which gave him an erroneous result of 78. What is the correct mean of the ten tests?
    A) 97.5   B) 70   C) 62.4   D) 52

> Multiply 78 by 8 to get the total points. Then divide by 10 to solve.

90) A bag contains 5 red balloons, 10 orange balloons, 8 green balloons, and 12 purple balloons. If a balloon is drawn from the bag at random, what is the probability that it will be orange?

 A) $\frac{2}{7}$ B) $\frac{1}{4}$ C) $\frac{1}{10}$ D) $\frac{1}{35}$

> This is a question on calculating basic probability. First of all, calculate how many items there are in total in the data set, which is also called the "sample space" or (S). Then reduce the data set if further items are removed. Probability can be expressed as a fraction. The number of items available in the total data set goes in the denominator. The chance of the desired outcome, which is also referred to as the event or (E), goes in the numerator of the fraction. You can determine the chance of the event by calculating how many items are available for the desired outcome.

91) A deck of cards contains 13 hearts, 13 diamonds, 13 clubs, and 13 spades. Cards are selected from the deck at random. Once selected, the cards are discarded and are not placed back into the deck. Two spades, one heart, and a club are drawn from the deck. What is the probability that the next card drawn from the deck will be a heart?

 A) $1/13$ B) $1/12$ C) $13/52$ D) $1/4$

> Reduce the sample space by the number of cards that have already been drawn for the denominator. Then determine how many hearts are left for the numerator.

92) Sam rolls a fair pair of six-sided dice. One of the die is black and the other is red. Each die has values from 1 to 6. What is the probability that Sam will roll a 4 on the red die and a 5 on the black die?

 A) $1/36$ B) $2/36$ C) $1/12$ D) $2/12$

> Determine how many combinations are possible on a set of dice. You may want to write down the possible combinations, which will make the answer clearer to you.

93) An owner of a carnival attraction draws teddy bears out of a bag at random to give to prize winners. She has 10 brown teddy bears, 8 white teddy bears, 4 black teddy bears, and 2 pink teddy bears when she opens the attraction at the start of the day. The first prize winner of the day receives a brown teddy bear. What is the probability that the second prize winner will receive a pink teddy bear?

 A) $1/24$ B) $1/23$ C) $2/24$ D) $2/23$

> For the denominator, reduce the sample space by the number of bears that have already been drawn. Then determine how many pink teddy bears are left for the numerator.

For questions 94 to 100, study the charts, paying attention to the legends and labels on each one. Read each question carefully to be sure what calculation you need to do. Then perform the operations to solve.

Look at the bar chart below and answer questions 94 to 97.

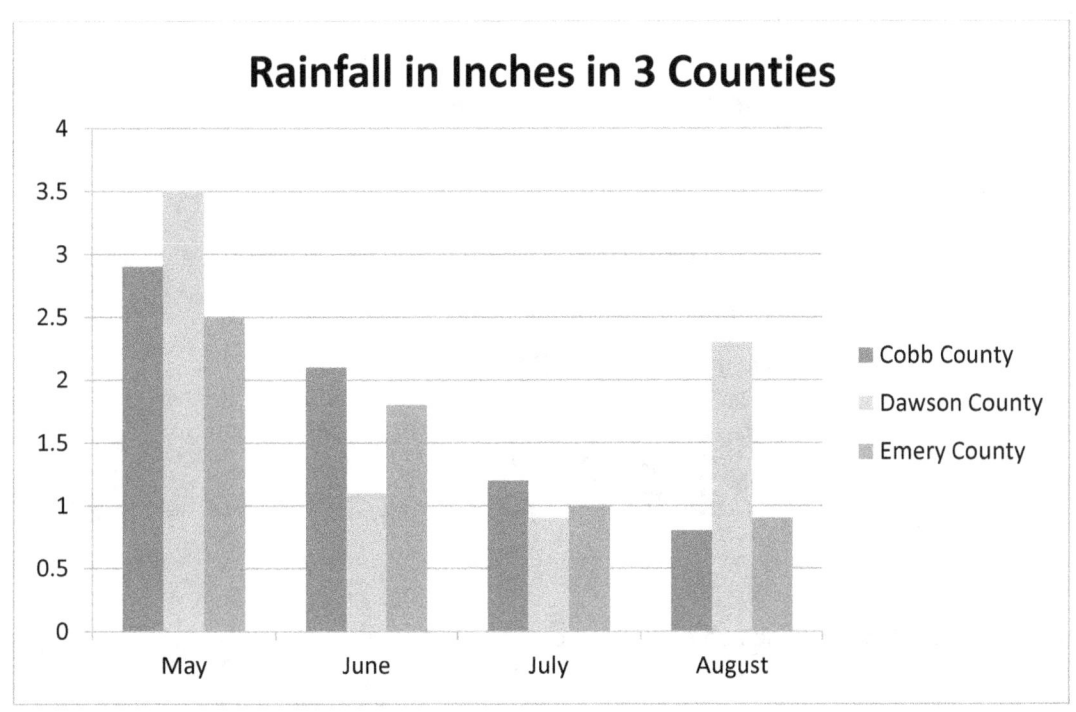

94) Approximately how many inches of rainfall did Cobb County have for July and August in total?
A) 0.7 inches B) 0.9 inches C) 2 inches D) 3.2 inches

95) What was the approximate difference in the amount of rainfall for Dawson County and Emery County for June?
A) Dawson County had 0.6 more inches of rainfall than Emery County.
B) Emery County had 0.6 more inches of rainfall than Dawson County.
C) Dawson County had 1.1 fewer inches of rainfall than Emery County.
D) Emery County had 1.1 fewer inches of rainfall than Dawson County.

96) What was the approximate total rainfall for Emery County for all four months?
A) 6.2 inches B) 6.8 inches C) 7.0 inches D) 7.4 inches

97) Which figure below best represents the total amount of rainfall in inches for the county that had the least amount of rainfall for all four months in total?
A) 6.2 B) 6.9 C) 7 D) 7.8

Look at the pie chart and information below and answer questions 98 to 100.

A zoo has reptiles, birds, quadrupeds, and fish. At the start of the year, they have a total of 1,500 creatures living in the zoo. The pie chart below shows percentages by category for the 1,500 creatures at the start of the year. At the end of the year, the zoo still has 1,500 creatures, but reptiles constitute 40%, quadrupeds 21%, and fish 16%.

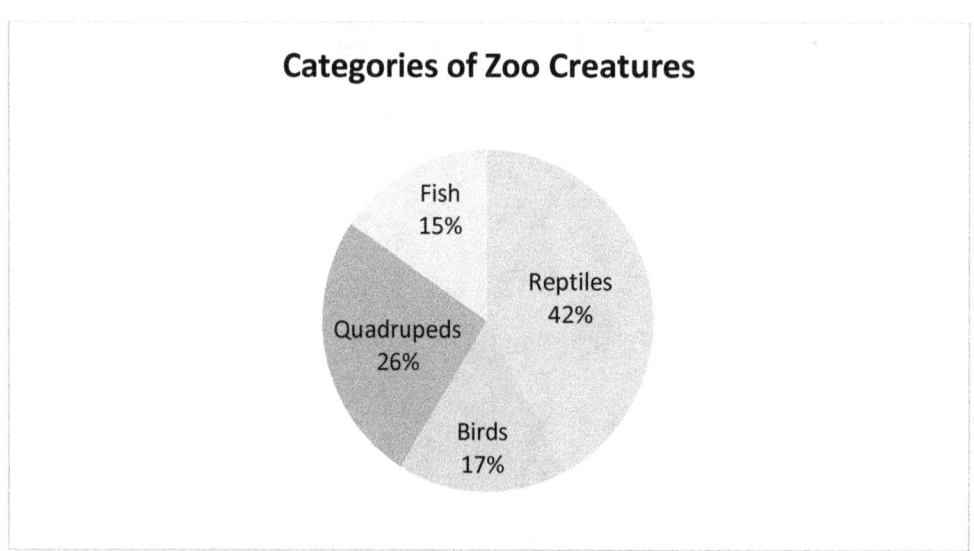

98) How many reptiles are in the zoo at the start of the year?
A) 225 B) 255 C) 390 D) 630

99) What was the difference between the number of quadrupeds at the start of the year and the number of fish at the start of the year?
A) There were 165 more fish than quadrupeds.
B) There were 165 more quadrupeds than fish.
C) There were 75 more fish than quadrupeds.
D) There were 75 more quadrupeds than fish.

100) What can be said about the number of birds at the end of the year when compared to the number of birds at the beginning of the year?
A) There were 23 more birds at the end of the year than at the beginning of the year.
B) There were 23 fewer birds at the end of the year than at the beginning of the year.
C) There were 90 more birds at the end of the year than at the beginning of the year.
D) There were 90 fewer birds at the end of the year than at the beginning of the year.

Algebra and Functions

Expressions with One Variable

101) Evaluate: $2x^2 - x + 5$ if $x = -2$
   A) 2 B) 7 C) 15 D) 17

> Step 1 – To perform the operations on the first term of the equation, multiply –2 by itself to square it. Then multiply this result by 2. Step 2 – To get your final answer, take the result from step 1 and subtract –2 and add 5.

102) Solve for x: $-6x + 5 = -19$
   A) 2 B) 4 C) 6 D) 8

> Isolate x to one side of the equation by subtracting 5 from both sides of the equation. Then multiply each side of the new equation by –6 to isolate x and solve.

103) If $4x - 3(x + 2) = -3$, then $x = ?$
   A) 9 B) 3 C) 1 D) –3

> Multiply the terms inside the parentheses by the –3 in front of the set of parentheses. Then simplify and isolate x to one side of the equation to solve.

104) If $\frac{3}{4}x - 2 = 4$, $x = ?$
   A) $\frac{8}{3}$ B) $\frac{1}{8}$ C) 8 D) –8

> Multiply each side of the equation by $\frac{4}{3}$ to get rid of the fraction. Then simplify the remaining new improper fraction and add the result of the simplified fraction to both sides of the equation solve.

105) What is the value of $\frac{x-3}{2-x}$ when $x = 1$?
   A) 2 B) –2 C) ½ D) –½

> Substitute 1 for the value of x. Then perform the subtraction in the numerator and the subtraction in the dominator. Then simplify the resulting fraction to solve.

Expressions with Two Variables

106) $x^2 + xy - y = 41$ and $x = 5$. What is the value of y?
   A) 2.6 B) 4 C) 6 D) –4

107) $x^2 + xy - y = 254$ and $x = 12$. What is the value of y?
   A) 110 B) 10 C) 11 D) 12

> For the two previous questions, substitute the stated values of x. Then perform the necessary operations on both sides of the equation to isolate y and solve.

Roots and Radicals

108) If $6 + 8(2\sqrt{x} + 4) = 62$, then $\sqrt{x} = ?$

   A) 3.25   B) 24   C) $\frac{3}{2}$ D) $\frac{2}{3}$

> Perform the multiplication on the parenthetical first. The get rid of the integers by subtracting them from both sides of the equations. Then divide by 16 to isolate $\sqrt{x}$ to solve.

109) Which of the answers below is equal to the following radical expression? $\sqrt{50}$

   A) $1 \div 50$ B) $2\sqrt{25}$ C) $2\sqrt{5}$ D) $5\sqrt{2}$

> Step 1 – Factor the number inside the square root sign. Step 2 – Look to see if any of the factors are perfect squares. In this case, the only factor that is a perfect square is 25. Step 3 – Find the square root of 25 then simplify.

110) $\sqrt{36} + 4\sqrt{72} - 2\sqrt{144} = ?$

   A) $2\sqrt{36}$ B) $2\sqrt{252}$ C) $18 + 24\sqrt{2}$ D) $-18 + 24\sqrt{2}$

> Step 1 – Find the common factors that are perfect squares. Step 2 – Factor the amounts inside each of the radical signs and simplify.

111) $\sqrt{7} \times \sqrt{11} = ?$

   A) $\sqrt{77}$ B) $\sqrt{18}$ C) $7\sqrt{11}$ D) $11\sqrt{7}$

> Step 1 – Multiply the numbers inside the radical signs. Step 2 – Put this product inside a radical symbol for your answer.

112) Simplify: $\sqrt{15} + 3\sqrt{15}$

   A) 45   B) $4\sqrt{15}$ C) $2\sqrt{15}$ D) $3\sqrt{30}$

> You can place the number 1 in front of the first radical because it will count only one time. Then add the numbers in front of the radical signs to solve.

113) Express as a rational number: $\sqrt[3]{\dfrac{216}{27}}$

    A) 3   B) 2   C) $\dfrac{7}{3}$ D) $\sqrt[3]{2}$

Step 1 – Find the cube roots of the numerator and denominator to eliminate the radical. Step 2 – Simplify further if possible. The cube root is a number that equals the required product when multiplied by itself two times.

Exponent Laws

114) $7^5 \times 7^3 = ?$

    A) 7^8 B) 7^{15} C) 14^8 D) 49^8

If the base number is the same, you need to add the exponents when multiplying, but keep the base number the same as before.

115) $xy^6 \div xy^3 = ?$

    A) xy^{18} B) xy^3 C) x^2y^3 D) xy^2

If the base number is the same, you need to subtract the exponents when dividing, but keep the base number the same as before.

116) $\sqrt{8x^4} \cdot \sqrt{32x^6} = ?$

    A) $8\sqrt{32x^{10}}$ B) $16x^{10}$ C) $16x^5$ D) $256x^{10}$

This question combines the laws or radicals with the laws of exponents.

117) A rocket flies at a speed of 1.7×10^5 miles per hour for 2×10^{-1} hours. How far has this rocket gone?

    A) 340,000 miles   B) 34,000 miles   C) 3,400 miles   D) 340 miles

Step 1: Add the exponents to multiply the 10's. Step 2: Multiply the miles per hour by the number of hours to get the distance traveled. Step 3: Then multiply these two results together to solve the problem.

118) $\sqrt{x^{\frac{5}{7}}} = ?$

A) $\frac{5x}{7}$
B) $\left(\sqrt[5]{x}\right)^7$
C) $\left(7\sqrt{x}\right)^5$
D) $\left(\sqrt[7]{x}\right)^5$

Step 1: Put the base number inside the radical sign. Step 2: The denominator of the exponent is the nth root of the radical. Step 3: The numerator is the new exponent.

119) $x^{-5} = ?$

A) $\frac{1}{x^{-5}}$
B) $\frac{1}{x^5}$
C) $-5x$
D) $\frac{1}{5x}$

120) $(-4)^{-3} = ?$

A) -64
B) $-\frac{1}{64}$
C) $\frac{1}{64}$
D) 64

Step 1: When you have an exponent that is a negative number, you need to set up a fraction, where 1 is the numerator. Step 2: Put the term with the exponent in the denominator, but remove the negative sign on the exponent.

121) $62^0 = ?$

A) -62
B) 0
C) 1
D) 62

122) $(25x)^0 = ?$

A) 0
B) 5
C) 1
D) 25

Any non-zero number raised to the power of zero is equal to 1.

Simplifying Rational Algebraic Expressions

123) $\dfrac{b + \frac{2}{7}}{\frac{1}{b}} = ?$

A) $b^2 + \frac{7}{2}$
B) $2b + \frac{7}{2}$
C) $b^2 + \frac{2b}{7}$
D) $\dfrac{b}{b + \frac{2}{7}}$

Step 1 – When the expression has fractions in both the numerator and denominator, treat the line in the main fraction as the division symbol. Step 2 – Invert the fraction that was in the denominator and multiply.

124) $\dfrac{x^2}{x^2+2x} + \dfrac{8}{x} = ?$

A) $\dfrac{x+8x+16}{x^2+2x}$ B) $\dfrac{x^2+8}{x^2+3x}$ C) $\dfrac{8x^2+16x}{x^3}$ D) $\dfrac{x^2+8x+16}{x^2+2x}$

Step 1 – Find the lowest common denominator. Since x is common to both denominators, we can convert the denominator of the second fraction to the LCD by multiplying the numerator and denominator of the second fraction by $(x+2)$. Step 2 – When both fractions have the LCD, add the numerators to solve.

125) Perform the operation and simplify: $\dfrac{2a^3}{7} \times \dfrac{3}{a^2} = ?$

A) $\dfrac{6a}{7}$ B) $\dfrac{5a^3}{7a^2}$ C) $\dfrac{2a^6}{21}$ D) $\dfrac{21}{2a^6}$

Step 1 – Multiply the numerator of the first fraction by the numerator of the second fraction to get the new numerator. Step 2 – Then multiply the denominators. Step 3 – Factor out a^2. Step 4 – Simplify.

126) $\dfrac{8x+8}{x^4} \div \dfrac{5x+5}{x^2} = ?$

A) $\dfrac{5x^2}{8}$ B) $\dfrac{8}{5x^2}$ C) $\dfrac{3x+3}{x^2}$ D) $\dfrac{x^2+8x+8}{x^4+5x+5}$

Step 1 – Invert and multiply by the second fraction. Step 2 – Cancel out $(x+1)$. Step 3 – Cancel out x^2.

Factoring Polynomials

127) Factor: $9x^3 - 3x$
A) $3x(3x^2-1)$ B) $3x(3x-1)$ C) $3x(x^2-1)$ D) $3x(x-3)$

128) Which of the following is a factor of: $2xy - 6x^2y + 4x^2y^2$
A) $(1+3x-2xy)$ B) $(1-3x+2xy)$ C) $(1+3x+2xy)$ D) $(1-3x-2xy)$

To factor an equation, you need to find the common factor for all of the terms in the equation. So, for the first question above, you need to divide all of the terms by $3x$. For the second question above, you need to divide all of the terms by $2xy$.

Equivalent Expressions

129) Which of the following mathematical expressions equals $3/xy$?

A) $3/x \times 3/y$ B) $3 \div 3xy$ C) $3 \div (xy)$ D) $1/3 \div 3xy$

> To find the equivalent expression, remember that the line in a fraction can be treated as the division symbol.

130) Which of the following is equivalent to the expression $2(x + 2)(x - 3)$ for all values of x?

A) $2x^2 - 2x - 12$ B) $2x^2 - 10x - 6$ C) $2x^2 + 2x - 12$ D) $2x^2 + 10x - 6$

> Perform the FOIL method on the parentheticals. Then multiply this result by 2 to solve. If you do not know how to perform the FOIL method, you may want to look at questions 132 and 133 first.

131) Which of the following is equivalent to $\dfrac{x}{5} \div \dfrac{9}{y}$?

A) $\dfrac{xy}{45}$ B) $\dfrac{9x}{5y}$ C) $\dfrac{1}{9} \times \dfrac{x}{5y}$ D) $\dfrac{1}{5} \times \dfrac{9}{5y}$

> Remember that to divide by a fraction, you need to invert the second fraction and the multiply.

Expanding Polynomials

132) Which of the following expressions is equivalent to $(x + 4y)^2$?

A) $2(x + 8y)$

B) $2x + 8y$

C) $x^2 + 8xy^2 + 16y^2$

D) $x^2 + 8xy + 16y^2$

133) $\left(2 + \sqrt{6}\right)^2 = ?$

A) 8

B) $8 + 2\sqrt{6}$

C) $8 + 4\sqrt{6}$

D) $10 + 4\sqrt{6}$

> When expanding polynomials, you should use the FOIL method: First – Outside – Inside – Last.
> We can demonstrate the FOIL method on an example equation as follows:
> $(a + b)(c + d) =$
> $(a \times c) + (a \times d) + (b \times c) + (b \times d) =$
> $ac + ad + bc + bd$

Linear Equations

134) A student has noticed that the more she studies, the better her grades are. Which of the following graphs illustrates this relationship?

A)

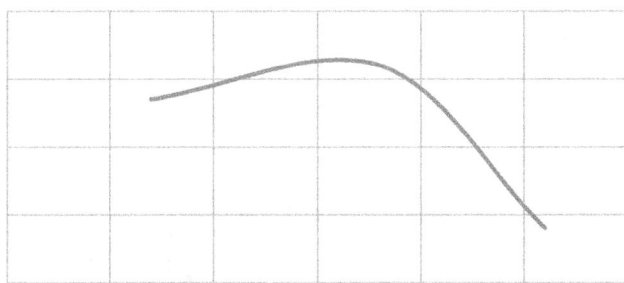

B)

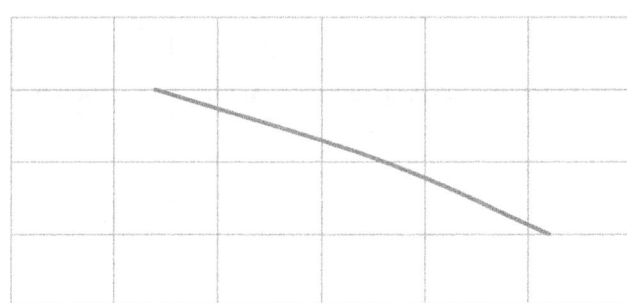

C)

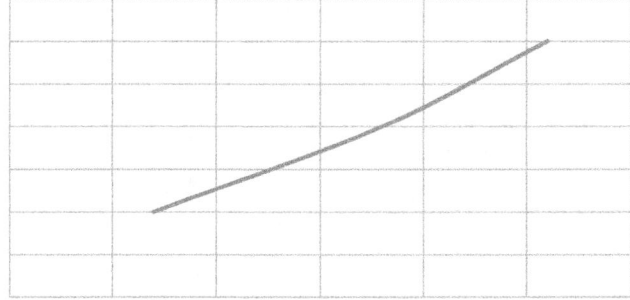

D)

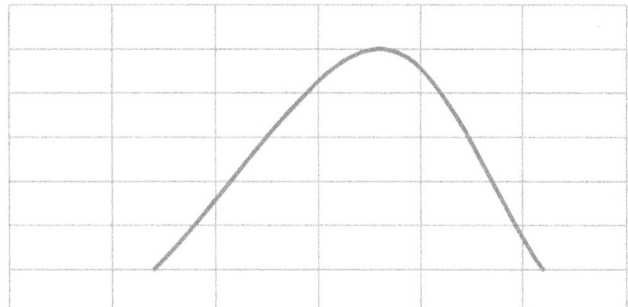

You will need to know the difference between positive linear relationships and negative linear relationships for the exam. In a positive linear relationship, an increase in one variable causes an increase in the other variable, meaning that the line will point upwards from left to right.

In a negative linear relationship, an increase in one variable causes a decrease in the other variable, meaning that the line will point downwards from left to right.

Please go on to the next page.

Algebraic Functions

135) The graph of a linear equation is shown below. Which one of the tables of values best represents the points on the graph?

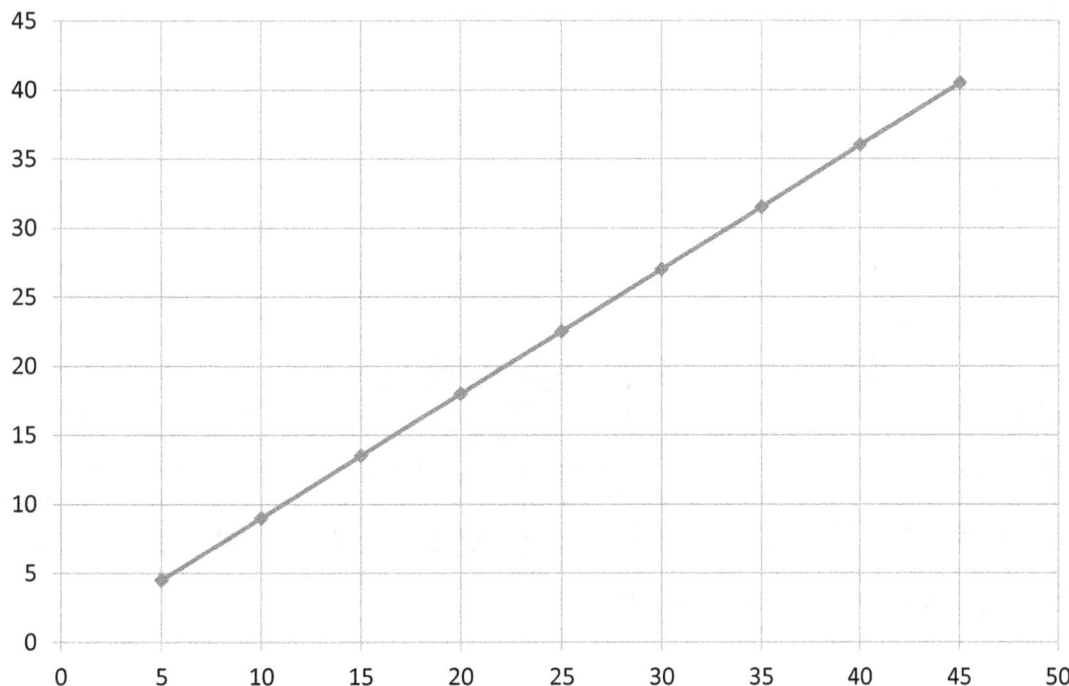

A)

x	y
5	5
10	10
15	15
20	20

B)

x	y
5	4
10	8
15	12
20	16

C)

x	y
5	4.5
10	9
15	13.5
20	18

D)

x	y
5	9
10	13
15	15
20	20

This is an example of an exam question involving algebraic functions. A function expresses the mathematical relationship between x and y. So, a certain recurring mathematical operation on x will yield the output of y. Step 1: Look carefully at the point that is furthest to the left on the graph. You will be able to eliminate several of the answer choices because they will not state this first coordinate correctly. Step 2: Try to work out the relationship between the coordinates of the first point to those of the next point on the line. Use the horizontal and vertical grid lines on the graph to help you.

136) What is the value of $f_1(2)$ where $f_1(x) = 5^x$?

A) 2^5 B) 10   C) 25   D) 25^2

For this type of algebraic function, substitute the value of 2 for x in the second expression.

137) For the two functions $f_1(x)$ and $f_2(x)$, tables of vales are given below. What is the value of $f_2(f_1(2))$?

x	$f_1(x)$
1	3
2	5
3	7
4	9
5	11

x	$f_2(x)$
2	4
3	9
4	16
5	25
6	36

A) 4

B) 5
C) 9
D) 25

Solve the first function for the value of 2. The take the resulting number and put it in as the value of x in the second function.

138) For the functions $f_2(x)$ listed below, x and y are integers greater than 1. If $f_1(x) = x^2$, which of the functions below has the greatest value for $f_1(f_2(x))$?

A) $f_2(x) = x/y$ B) $f_2(x) = y/x$ C) $f_2(x) = xy$ D) $f_2(x) = x - y$

Look at all of the answer choices, and substitute any positive integers for x and y. Then try the same using negative integers.

139) If $f(x) = x^2 + 3x - 8$, what is $f(x + 3)$?

A) $(x + 3)^2 + 3x - 8$
B) $(x + 3)^2 + 3(x + 3) - 8$
C) $x^2 + 3x - 5$
D) $3(x^2 + 3x - 8)$

In the first equation, substitute $x + 3$ for x to solve.

Logarithmic Functions

140) If $\log_3(x + 2) = 4$, then $x = $?
A) 66
B) 79
C) 81
D) 83

To convert a logarithmic function to an exponent, the number after the equals sign (4 in this problem) becomes the exponent. The small subscript number after "log" (3 in this problem) becomes the base number. Then perform the multiplication to solve. $\log_y Z = x$ is always the same as: $y^x = Z$

Quadratic Equations

141) Simplify: $(x - y)(x + y)$
A) $x^2 - 2xy - y^2$ B) $x^2 + 2xy - y^2$ C) $x^2 + y^2$ D) $x^2 - y^2$

Use the FOIL method on quadratic equations like this one when the instructions tell you to simplify. If you do not remember how to perform the FOIL method, look at questions 132 and 133 again. Then try the next question.

142) $(3x + y)(x - 5y) = $?
A) $3x^2 - 14xy - 5y^2$
B) $3x^2 - 14xy + 5y^2$
C) $3x^2 + 14xy - 5y^2$
D) $3x^2 + 14xy + 5y^2$

Linear Inequalities

143) $50 - \dfrac{3x}{5} \geq 41$, then $x \leq$?

A) 15   B) 25   C) 41   D) 50

> Step 1: Isolate the whole numbers to one side of the inequality. Step 2: Get rid of the fraction by multiplying each side by 5. Step 3: Divide to simplify further. Step 4: Isolate the variable to solve.

144) The cost of one toy is equal to y. If $x - 2 > 5$ and $y = x - 2$, then the cost of 2 toys is greater than which one of the following?

A) $x - 2$ B) $x - 5$ C) $y + 5$ D) 10

> Look to see if the inequality and the equation have any variables or terms in common. In this problem, both the inequality and the equation contain $x - 2$. The cost of one toy is represented by y, and y is equal to $x - 2$. So, we can substitute values from the equation to the inequality.

Quadratic Inequalities

145) Solve for x: $x^2 - 9 < 0$
A) $x < -3$ or $x > 3$
B) $x > -3$ or $x < 3$
C) $x < -3$ or $x < 3$
D) $x > -3$ or $x > 3$

> For quadratic inequality problems like this one, you need to factor the inequality first. We know that the factors of −9 are: −1 × 9; −3 × 3; 1 × −9. We do not have a term with only the x variable, so we need factors that add up to zero. −3 + 3 = 0. So, try to solve the problem based on these facts. Be sure to check your answer by substituting greater or lesser values (like 4 and −4) into the original inequality.

146) Solve for x: $x^2 - 5x + 6 \leq 0$
A) $2 \geq x \geq 3$
B) $2 \leq x \leq 3$
C) $x < -3$ or $x < 2$
D) $x > -2$ or $x > 3$

> Here is another quadratic inequality problem. Remember to factor the inequality first. We know that the factors of 6 are: 1 × 6 and 2 × 3. We have a term with the x variable, so we need factors that add up to five. 2 + 3 = 5. So, try to solve the problem based on these facts. Be sure to check your work to be sure the signs point the right way by substituting values into the original inequality.

Systems of Equations

147) What ordered pair is a solution to the following system of equations?
$x + y = 7$
$xy = 12$

A) (2, 6)   B) (6, 2)   C) (4, 2)   D) (3, 4)

Step 1: Look at the multiplication equation and find the factors of 12. Step 2: Add the factors in each set together to see if they equal 7 to solve the addition in the first equation.

148) Solve by elimination: $3x + 3y = 15$ and $x + 2y = 8$

A) $x = -18$ and $y = 13$
B) $x = -2$ and $y = 3$
C) $x = 2$ and $y = 3$
D) $x = 3$ and $y = 2$

Step 1: Look at the x term of the first equation, which is 3x. In order to eliminate the x variable, we need to multiply the second equation by 3. Step 2: Subtract this result from the first equation to solve.

Sequences and Series

149) What is the next number in the following sequence? 1, 5, 9, 13, 17, . . . **(M)**

A) 20
B) 21
C) 30
D) 40

Sequences are numbers in a list like the following: 1, 3, 5, 7, 9. In a series, the numbers are added: 1 + 3 + 5 + 7 + 9. In an arithmetic sequence, the difference between one number and the next is known as a constant. In other words, you add the same value each time until you reach the end of the sequence.
The formula for the nth number of an arithmetic sequence is a + [d × (n − 1)], where variable *a* represents the starting number and variable *d* represents the difference or constant.

150) What is the next number in the following sequence? 2, 6, 18, 54, . . .
A) 60
B) 72
C) 80
D) 162

When the sequence cannot be solved by addition, then you usually have a geometric sequence.

In a geometric sequence, each number is found by multiplying the previous term by a factor known as a common ratio. Where the first number is represented by variable *a* and the factor (called the "common ratio") is represented by variable *r*, the formula for calculating the nth item in a geometric sequence is: $ar^{(n-1)}$

Bonus Exercises – Solutions and Explanations

1) The correct answer is C. The problem is asking for the total for all three years, so we add the three figures together: $25,135 + $32,787 + $47,004 = $104,926

2) The correct answer is D. For questions that ask you to calculate the change given to a customer, you need to take the amount of money the customer gives the cashier and subtract the amount of the purchase: $50.00 – $41.28 = $8.72

3) The correct answer is D. Multiplication problems will often include the words 'each' or 'every.' The problem states that the salesperson earns a $175 referral fee on every customer, so the referral fee was earned 8 times this month. We need to multiply the amount of the referral fee by the number of customers to solve: $175 × 8 = $1400

4) The correct answer is C. Division problems will often include the word 'per.' The problem states that the employee works 30 hours per week. So, we divide the total weekly amount by the number of hours to solve: $535.50 ÷ 30 = $17.85

5) The correct answer is B. When you have to add a negative number to a positive number, you are really subtracting. So, add the business profits and subtract the business losses:
953 + 1502 – 286 – 107 = 2062

6) The correct answer is A. In this problem, we need to subtract the excess of the depth of Lake Bajo from the location below sea level of Lake Alto. The location below sea level of Lake Alto is a negative number, so we subtract as follows: –35 – 62 = –97. Remember to express your result as a negative number.

7) The correct answer is B. In order to express a fraction as a decimal, treat the line in the fraction as the division symbol: 3/5 = 3 ÷ 5 = 0.60. Be careful with the decimal placement in your final result.

8) The correct answer is C. To express a decimal number as a percent, move the decimal point two places to the right and add the percent sign: 0.55 = 55.0%

9) The correct answer is D. In order to express a fraction as a percentage, you need to divide and then express the result as a percentage. Step 1 – Treat the line in the fraction as the division symbol: 5/14 = 5 ÷ 14 = 0.357. Step 2 – To express the result from Step 1 as a percentage, we need to move the decimal point two places to the right and add the percent sign: 0.357 = 35.7%

10) The correct answer is D. For your exam, you should be able to recognize the equivalent fractions for commonly-used decimal numbers. If you are unsure, perform division on the answer choices to check: 3/4 = 3 ÷ 4 = 0.75

11) The correct answer is A. For your exam, you should be able to recognize the equivalent fractions for commonly-used percentages. If you are unsure, perform division on the answer choices to check: 1/3 = 1 ÷ 3 = 0.3333 = 33%

12) The correct answer is C. Any given percentage is out of 100%, so we divide by 100 to express a percentage as a decimal. So, move the decimal point two places to the left and remove the percent sign: 45% = 45 ÷ 100 = 0.45

13) The correct answer is B. Express both amounts as decimal numbers and multiply to solve:
14¼ pounds × 36 cents per pound = 14.25 × 0.36 = $5.13

14) The correct answer is C. There are 60 minutes in an hour, so multiply the minutes in the hour by the decimal number given in the problem to solve: 60 minutes × 0.35 hour = 60 × 0.35 = 21 minutes

15) The correct answer is A. Step 1 – Subtract the discount from the original price: $24 – $5 = $19. Step 2 – Take the result from Step 1 and multiply by the number of units sold: $19 × 12 = $228

16) The correct answer is D. Step 1 – Determine the total number of hours worked: 7 hours per day for 4 days = 7 × 4 = 28 hours. Step 2 – Calculate the profit the company makes per hour. The customer was billed $45 per hour for the employee's work, and he was paid $25 per hour: $45 – $25 = $20 profit per hour. Step 3 – Multiply the total number of hours by the profit per hour to solve: 28 hours × $20 profit per hour 28 × 20 = $560

17) The correct answer is A. Step 1 – Calculate how many minutes there are in 40 hours: 40 hours × 60 minutes per hour = 2400 minutes. Step 2 – Divide the amount of prescriptions into the previous result to get the rate: 2400 ÷ 250 = 9.6 minutes per prescription

18) The correct answer is C. Take the amount of orders that were delivered on time and divide by the amount of total orders: 105 ÷ 120 = 0.875 = 87.5%

19) The correct answer is B. On Monday cell growth was 27, and for all of the days Tuesday through Friday, cell attrition was 13 per day. Step 1 – Cell attrition is a negative number, so perform multiplication to get the total for the four days (Tuesday through Friday): –13 × 4 = –52. Step 2 – On Monday cell growth was 27, so add this to the result from Step 1 to solve: –52 + 27 = –25

20) The correct answer is B. To find the average, you need to find the total, and then divide the total by the number of hours. Step 1 – Find the total: 23 + 25 + 26 + 24 + 22 = 120. Step 2 – Divide the result from Step 1 by the number of hours: 120 ÷ 5 = 24

21) The correct answer is D. Step 1 – Take the 66 units of cement powder for the current batch and divide by the 3 units stated in the original ratio: 22 ÷ 3 = 22. Step 2 – Multiply by the 2 units of sand stated in the original ratio to get your answer: 2 × 22 = 44

22) The correct answer is D. The problem states that we are working with a ratio, so we can think of the employees and the supervisors as forming separate groups. Step 1 – Add the two groups together: 50 + 1 = 51. Step 2 – Take the total number of employees stated in the problem and divide this by the figure calculated in Step 1 to get the number of supervisors: 255 ÷ 51 = 5

23) The correct answer is D. The problem uses the phrase '2 out of every 20 employees' so we know that there are 2 employees who form a subset within each group of 20. Step 1 – Take the total number of employees and divide this by 20: 480 ÷ 20 = 24. Step 2 – Take the result from Step 1 and multiply by the amount in the subset to solve: 24 × 2 = 48

24) The correct answer is C. Step 1 – Calculate the amount of time spent on the initial job to do 3 wheel covers: 8:10 to 8:22 = 12 minutes. Step 2 – Calculate how many minutes are needed to change 1 wheel

cover: 12 minutes ÷ 3 = 4 minutes each. Step 3 – Divide the figure from Step 2 into 60 minutes to solve: 60 ÷ 4 = 15

25) The correct answer is C. Step 1 – Add the whole numbers. The whole numbers are the numbers in front of the fractions: 15 + 13 = 28. Step 2 – Add the fractions. If you have two fractions that have the same denominator, you add the numerators and keep the common denominator: 2/8 + 5/8 = 7/8. Step 3 – Combine the results from Step 1 and Step 2 to get your new mixed number to solve the problem: 28 + 7/8 = $28^{7}/_{8}$

26) The correct answer is A. Step 1 – Add the whole numbers: 2 + 4 = 6. Step 2 – Add the fractions. If you have two fractions that have the same denominator, you add the numerators and keep the common denominator: 1/8 + 3/8 = 4/8. Step 3 – Simplify the fraction from Step 2: 4/8 = (4 ÷ 4)/(8 ÷ 4) = 1/2. Step 4 – Combine the results from Step 1 and Step 3 to get your new mixed number to solve the problem: 6 + 1/2 = $6^{1}/_{2}$

27) The correct answer is A. Step 1 – Subtract the whole numbers: 5 – 4 = 1. Step 2 – Subtract the fractions. If you have two fractions that have the same denominator, you subtract the numerators and keep the common denominator: 3/16 – 1/16 = 2/16. Step 3 – Simplify the fraction from Step 2: 2/16 = (2 ÷ 2)/(16 ÷ 2) = 1/8. Step 4 – Combine the results from Step 1 and Step 3 to get your new mixed number to solve the problem: 1 + 1/8 = $1^{1}/_{8}$

28) The correct answer is B. Add the three figures together to solve: 0.25 + 0.50 + 0.10 = 0.85. Remember to be sure to put the decimal point in the correct place when you work out the solution to problems like this one.

29) The correct answer is C. Add the percentages together to solve: 25% + 50% = 75%

30) The correct answer is D. Step 1 – Multiply the whole numbers: 5 × 1 = 5. Step 2 – Multiply the whole number by the fraction: 5 × 1/4 = 5/4. Step 3 – Convert the fraction from Step 2 to a mixed number: 5/4 = $1^{1}/_{4}$. Step 4 – Combine the results from Step 1 and Step 3 to get your new mixed number: 5 + $1^{1}/_{4}$ = $6^{1}/_{4}$. Step 5 – Convert the result from Step 4 to hours and minutes: $6^{1}/_{4}$ hours = 6 hours and 15 minutes

31) The correct answer is B. Step 1 – Convert the first fraction to the common denominator: 1/8 = (1 × 4)/(8 × 4) = 4/32. Step 2 – Add one more increment to this to get your result: 4/32 + 1/32 = 5/32

32) The correct answer is A. Step 1 – Work out the cost for the first supplier: 50 units × $0.50 = $25. Step 2 – Compare to other deals to solve: The other deals are $27.50 and $30, so $25 is the best deal.

33) The correct answer is D. Step 1 – Determine the duration of the stay in weeks and nights: 9 nights = 1 week + 2 nights. Step 2 – Add the cost for 1 week to the cost for 2 days to solve: $280 + (2 × $45) = $280 + $90 = $370

34) The correct answer is D. Step 1 – Determine the dollar value of the discount: $15 – $12 = $3. Step 2 – Divide the result from Step 1 by the original price to get the percentage: $3 ÷ $15 = 0.20 = 20%

35) The correct answer is C. Step 1 – Determine the dollar value of the mark-up on the mug: $9 retail price – $3 cost = $6 mark-up. Step 2 – Calculate the percentage of the mark-up by dividing the dollar

value of the mark-up by the cost: $6 ÷ $3 = 2.00 = 200%. Step 3 – Use the percentage mark-up from the previous step to determine the dollar value of the mark-up on the bowl: $4 × 200% = $4 × 2 = $8. Step 4 – Add the dollar value of the mark-up for the bowl to the cost of the bowl to get the retail price: $8 + $4 = $12

36) The correct answer is D. To calculate a reverse percentage you need to divide, rather than multiply. So, take the $20 discount and divide by the 25% percentage: $20 ÷ 25% = $20 ÷ 0.25 = $80

37) The correct answer is C. Step 1 – Add the times for the first two processes and express in terms of hours and minutes: Production time of 3 hours and 25 minutes + Bottling and labeling time of 1 hour and 40 minutes = 3 hours + 1 hour + 25 minutes + 40 minutes = 4 hours and 65 minutes = 5 hours and 5 minutes. Step 2 – Add the time for the packaging process of 26 hours to the result from Step 1: 5 hours and 5 minutes + 26 hours = 31 hours and 5 minutes. Step 3 – Determine the time that the batch will be ready for shipment. 31 hours and 5 minutes have passed. In other words, a period of 24 hours and an additional 7 hours and 5 minutes have passed. The process started on Monday at 10:30 am, so by Tuesday at 10:30 am, 24 hours will have passed. An additional 7 hours and 5 minutes takes us to Tuesday at 5:35 pm.

38) The correct answer is C. Step 1 – Determine the cost from the first supplier: 240 × 0.25 = $60. The tax on this will be $60 × 6.5% = $60 × 0.065 = $3.90. Then add the tax to the cost to get the total: $60 + $3.90 = $63.90. Step 2 – Determine the total cost from the second supplier: $58 cost + ($58 × 0.065 tax) = $58 + 3.77 = $61.77. So, you will get the better deal from the second supplier at $61.77.

39) The correct answer is D. Step 1 – Determine how many days are needed to make the small frames. 20 small frames can be made in 4 days: 20 frames ÷ 4 days = 5 small frames per day. The customer wants 40 small frames, so divide by the rate to determine how many days are going to be needed for the small frames: 40 frames ÷ 5 per day = 8 days. Step 2 – Determine how many days are going to be needed to make the large frames. 21 larges frames can be made in 3 days: 21 ÷ 3 = 7 large frames per day. 64 large frames need to be made for the order: 64 ÷ 7 = 9.1 days. Step 3 – Add the results from the two previous steps to solve: 8 days + 9.1 days = 17.1 days, which we round down to 17 days.

40) The correct answer is C. Step 1 – Calculate the percentage of work completed per day. 12.5% of the work has been completed in 4 days: 12.5 % ÷ 4 days = 3.125% per day. Step 2 – Determine how many days in total are needed to complete the entire job by dividing 100% by the result from the previous step: 100% ÷ 3.125% = 32 days. Step 3 – Determine the number of days remaining: 32 days in total – 4 days completed = 28 days remaining

41) The correct answer is B. From the formula, we can see that 1 foot = 0.3048 meters. To solve, multiply the amount of 538 feet, stated in the question, by 0.3048: 538 × 0.3048 = 163.98, which we round up to 164.

42) The correct answer is D. Step 1 – Add the feet together: 123 + 138 = 261 feet. Step 2 – Add the inches together: 6 + 8 = 14 inches. Step 3 – Convert the inches to feet and inches if the result from Step 2 is 12 inches or more: 14 inches = 1 foot 2 inches. Step 4 – Combine the results from Step 1 and Step 3 to solve: 261 feet + 1 foot 2 inches = 262 feet 2 inches

43) The correct answer is A. Step 1 – Convert the weight of the full box from pounds and ounces to just ounces. We are using the formula 1 pound = 16 ounces, so 8 pounds and 5 ounces = (8 × 16) + 5 = 128 + 5 = 133 ounces. Step 2 – The problem states that the box weighs 7 ounces when it is empty. So,

subtract the weight of the empty box from the weight of the full box to get the weight of the product inside the box: 133 ounces − 7 ounces = 126 ounces. Step 3 − The problem tells us that each supplement weighs 0.75 ounces. Take the total weight from the previous step and divide by the weight per unit to determine how many units the box contains: 126 ounces ÷ 0.75 ounces = 168 units

44) The correct answer is B. Step 1 − Convert the mixed numbers to decimals and then multiply: 50¼ feet × 60¼ feet = 50.25 × 60.25 = 3027.5625 square feet. Step 2 − The price is given in square yards, so convert the square feet from the previous step to square yards. The formula states that 1 square yard = 9 square feet, so 1/9 square yard = 1 square foot: 3027.5625 square feet ÷ 9 = 336.3958 square yards. Step 3 − Calculate the cost: 336.3958 × $5.25 = $1765.92, which we round to $1,766.

45) The correct answer is B. Step 1 − Calculate the amount of remaining stock in inches: (2 × 75 inches) + (4 × 25.25 inches) = 150 + 101 = 251 inches. Step 2 − Convert the existing stock from inches to yards: 1 foot = 12 inches and 1 yard = 3 feet, so there are 36 inches in 1 yard. So, divide the amount of inches by 36 to convert to yards: 251 ÷ 36 = 6.97 yards. Step 3 − Calculate the amount required to restock. 60 yards are required in total, and there are 6.97 yards on hand, so subtract to find out how many more yards are needed to get the stock back up to 60 yards: 60 − 6.97 = 53.03 yards needed. Step 4 − The yarn comes in 5-yard balls, so calculate how many balls to buy to cover the 53.03 yards that are required: 53.03 ÷ 5 = 10.6 balls. It is not possible to buy a fractional part of a ball, so we round up to 11 balls.

46) The correct answer is D. Step 1 − Convert 0.75 grams to milligrams. 1 gram = 1,000 milligrams, so 0.75 grams × 1,000 = 750 milligrams. Step 2 − The normal ratio is in the amount of 50 milligrams, so divide the result from the previous step by 50: 750 ÷ 50 = 15. So, 15 times more active ingredient is being used than normal. Step 3 − Determine the amount of liquid. Since 15 times more of the active ingredient is being used, we also need to use 15 times more of the liquid: 1.5 milliliters × 15 = 22.5 milliliters

47) The correct answer is A. Our points are (5, 2) and (7, 4), so substitute the values into the midpoint formula.
$(x_1 + x_2) ÷ 2$, $(y_1 + y_2) ÷ 2$
$(5 + 7) ÷ 2$ = midpoint x, $(2 + 4) ÷ 2$ = midpoint y
$12 ÷ 2$ = midpoint x, $6 ÷ 2$ = midpoint y
6 = midpoint x, 3 = midpoint y

48) The correct answer is B. First, find the midpoint of the x coordinates for (**−4**, 2) and (**8**,−6).

midpoint $x = (x_1 + x_2) ÷ 2$

midpoint $x = (−4 + 8) ÷ 2$

midpoint $x = 4 ÷ 2$

midpoint $x = 2$

Then find the midpoint of the y coordinates for (−4, **2**) and (8,**−6**).

midpoint $y = (y_1 + y_2) \div 2$

midpoint $y = (2 + -6) \div 2$

midpoint $y = -4 \div 2$

midpoint $y = -2$

So, the midpoint is (2, −2)

49) The correct answer is D. Substitute the values (2, 3) and (6, 7) into the formula.
$$d = \sqrt{(x_2 - x_1)^2 + (y_2 - y_1)^2}$$
$$d = \sqrt{(6 - 2)^2 + (7 - 3)^2}$$
$$d = \sqrt{4^2 + 4^2}$$
$$d = \sqrt{16 + 16}$$
$$d = \sqrt{32}$$

50) The correct answer is A. Substitute the values into the slope-intercept formula.

$y = mx + b$

$315 = m5 + 15$

$315 - 15 = m5 + 15 - 15$

$300 = m5$

$300 \div 5 = m5 \div 5$

$60 = m$

51) The correct answer is A. As y increases by 5, x decreases by 5. So, the slope is −1. The line includes point (20, 15), which is the fifth point from the left.

52) The correct answer is A. Remember that the y intercept is where the line crosses the y axis, so $x = 0$ for the y intercept. Begin by substituting 0 for x.

$y = x + 14$
$y = 0 + 14$
$y = 14$
Therefore, the coordinates (0, 14) represent the y intercept.

On the other hand, the x intercept exists where the line crosses the x axis, so y = 0 for the x intercept. Now substitute 0 for y.

y = x + 14
0 = x + 14
0 − 14 = x + 14 − 14
−14 = x

So, the coordinates (−14, 0) represent the x intercept.

53) The correct answer is A. The x intercept is the point at which a line crosses the x axis of a graph. In order for the line to cross the x axis, y must be equal to zero at that particular point of the graph. On the other hand, the y intercept is the point at which the line crosses the y axis. So, in order for the line to cross the y axis, x must be equal to zero at that particular point of the graph. First, substitute 0 for y in order to find the x intercept.

$x^2 + 2y^2 = 144$

$x^2 + (2 \times 0) = 144$

$x^2 + 0 = 144$

$x^2 = 144$

$x = 12$

Then substitute 0 for x in order to find the y intercept.

$x^2 + 2y^2 = 144$

$(0 \times 0) + 2y^2 = 144$

$0 + 2y^2 = 144$

$2y^2 \div 2 = 144 \div 2$

$y^2 = 72$

$y = \sqrt{72}$

So, the y intercept is $(0, \sqrt{72})$ and the x intercept is (12, 0).

54) The correct answer is B. From the formula, we can see that the area of a triangle is ½ (base × height). So, substitute the values to solve: ½ (base × height) = ½ (12 × 14) = ½ × 168 = 84 square inches

55) The correct answer is C. Use the Pythagorean Theorem to solve. $C = \sqrt{A^2 + B}$

$$C = \sqrt{A^2 + B^2} = \sqrt{3^2 + 2^2} = \sqrt{9 + 4} = \sqrt{13}$$

56) The correct answer is C. The sum of the angles in a triangle is 180 degrees. So, subtract the measurements of the other two angles to solve: 180° − 47° − 44° = 89°

57) The correct answer is D. From the tip after the question, we can see that a circle has 360 degrees. So, subtract to solve: 360 − 82 − 79 − 46 − 85 = 68

58) The correct answer is A. From the formula, we can see that the area of a circle ≈ 3.14 × (radius)². So, put in 12 feet for the radius to solve: 3.14 × (12 × 12) = 3.14 × 144 = 452.16

59) The correct answer is C. From the formula, we know that the circumference of a circle ≈ 3.14 × diameter. The problem states that the diameter of the tractor tire is 46.5 inches, so use that in the formula to solve: 3.14 × 46.5 = 146.01 inches

60) The correct answer is D. The area of a rectangle = length × width. Your quilt is 6 feet long and 5 feet wide, so multiply to solve: 6 × 5 = 30

61) The correct answer is B. The perimeter of a rectangle = 2(length + width). Your field is 12 yards long and 9 yards wide, so use the formula to solve: 2(12 + 9) = 2 × 21 = 42

62) The correct answer is B. Step 1 – The area of a circle ≈ 3.14 × radius². Here, we are given the area, so we have to divide by 3.14, instead of multiplying by 3.14, as stated in the formula: 78.5 ÷ 3.14 = 25. Step 2 – The result from the previous step is the radius squared. A squared number is the result of a number that has been multiplied by itself. 5 × 5 = 25, so the length of the radius of the pond is 5 feet. Step 3 – Remember that diameter is double the radius, so if the radius is 5, the diameter is 10 feet.

63) The correct answer is A. For questions on rearranging formulas like this one and the previous one, it is very likely that you are going to have to divide the largest number in the question by a smaller number in order to solve the problem. From the formula, we know that the area of a rectangle = length × width. Here, we are given the area (the larger number of 360), so we need to divide that by the length (the smaller number of 30 feet) in order to get the width: 360 ÷ 30 = 12 feet

64) The correct answer is D. The volume of a rectangular solid = length × width × height. The tank is 5 feet wide, 8 feet long, and 3 feet high, so multiply to solve: 5 × 8 × 3 = 120

65) The correct answer is A. A cube is a three-dimensional object in which all sides have the same length. The volume of a cube = side length³. So, put the length of the side in the formula to solve: 18 × 18 × 18 = 5832

66) The correct answer is A. Step 1 – Calculate in cubic inches the volume of the sphere when it is full. The tank is 72 inches across on the inside, so the radius is 36 inches. The volume of a sphere ≈ 4/3 ×

$3.14 \times radius^3$: $4/3 \times 3.14 \times 36^3$ = 195,333.12 cubic inches. Step 2 – Calculate in cubic inches how much milk remains in the sphere. The tank is now 80% full of milk: 195,333.12 cubic inches × 0.80 = 156,266.50 cubic inches, which we round to 156,267 cubic inches.

67) The correct answer is B. The volume of a cylinder ≈ $3.14 \times height \times radius^2$. Your tank has a 5 meter radius and is 21 meters in height: $3.14 \times 21 \times 5^2$ = 3.14 × 21 × 25 = 1648.50 cubic meters

68) The correct answer is C. Step 1 – Calculate the volume of the large cone. The large cones are 6 inches high and have a 1.5 inch radius. The volume of a cone ≈ ($3.14 \times height \times radius^2$) ÷ 3 = (3.14 × 6 × 1.5 × 1.5) ÷ 3 = 14.13. Step 2 – Calculate the volume of the medium cone. The medium cones are 5 inches high and have a 1 inch radius: ($3.14 \times height \times radius^2$) ÷ 3 = (3.14 × 5 × 1 × 1) ÷ 3 = 5.23. Step 3 – Calculate the difference between the volume of the two cones: 14.13 – 5.23 = 8.90

69) The correct answer is D. Step 1 – Calculate the dimensions of the floor in inches: 8 feet × 12 inches per foot = 96 inches long; 4 feet × 12 inches in a foot = 48 inches wide. Step 2 – Determine how many wooden pieces will fit along the length of the floor. If we lay the 12-inch side of the wooden piece against the length of the room, we can lay 8 of these side by side to cover the 96-inch length: 96 ÷ 12 = 8. Step 3 – Determine how many wooden pieces can fit along the width. 48-inch-wide floor ÷ 6-inch-wide pieces = 48 ÷ 6 = 8 pieces. Step 4 – Multiply the results from steps 2 and 3 to get the total number of pieces needed for the job: 8 × 8 = 64

70) The correct answer is C. Area of a rectangle = length × width. The wall is 16 feet by 11 feet, so multiply to solve: 16 × 11 = 176

71) The correct answer is A. We know that the volume of a rectangular solid = length × width × height. Here, we are given the volume, so we need to divide that by the length and then the width in order to find the height: (1080 ÷ 12) ÷ 9 = 90 ÷ 9 = 10 feet

72) The correct answer is B. Step 1 – Find the volume in terms of cubic inches. Remember that radius is half of diameter. Here we have a diameter of 12, so the radius is 6. Cylinder volume ≈ $3.14 \times radius^2 \times height$ ≈ $3.14 \times 6^2 \times 18$ ≈ 3.14 × 36 × 18 ≈ 2034.72. Step 2 – Convert the volume in cubic inches to gallons. 1 gallon = 231 cubic inches, so divide by 231 to convert to gallons: 2034.72 ÷ 231 = 8.8 gallons

73) The correct answer is D. Step 1 – First we need to calculate the volume in terms of cubic feet. The volume of a cube = $(length\ of\ side)^3$. The length of the side is 9 feet, so the volume is 9 × 9 × 9 = 729 cubic feet. Step 2 – We have to convert the result from Step 1 to cubic inches. From the formula, we can see that 1 cubic foot = 1,728 cubic inches, so multiply to solve: 729 × 1,728 = 1,259,712 cubic inches

74) The correct answer is A. Step 1 – Calculate in cubic feet the volume of the container when it is full. The container is 25 feet long, 12 feet wide and 18 feet high. To find the volume of a rectangular solid, we use the formula: length × width × height = 25 × 12 × 18 = 5,400 cubic feet. Step 2 – Calculate in cubic feet how much product is in the container. The container is now 75% full: 5,400 cubic feet × 0.75 = 4,050 cubic feet. Step 3 – Convert the cubic feet to yards. 1 cubic yard = 27 cubic feet. The formula is cubic yards to cubic feet, but you are converting from cubic feet to cubic yards, so you need to divide: 4,050 cubic feet ÷ 27 = 150 cubic yards

75) The correct answer is C. Step 1 – Calculate the amount of remaining stock in quarts and ounces: [2 × (16 cups and 7 ounces)] + [3 × (20 cups and 4 ounces)] = 32 cups and 14 ounces + 60 cups and 12 ounces = 92 cups and 26 ounces. Step 2 – Convert the existing stock from cups to quarts: 1 quart = 4

cups, so divide the amount of cups by 4 to convert to quarts: (92 cups ÷ 4) + 26 ounces = 23 quarts and 26 ounces. There are 32 ounces in a quart, so we cannot convert the remaining 26 ounces to quarts. Step 3 – Calculate the amount required to restock. 50 quarts are required in total, and you have approximately 23 quarts on hand, so subtract to find out how many more quarts you need to get the stock back up to 50 quarts: 50 – 23 = 27 quarts needed. Step 4 – The chemical comes in 5-quart containers, so calculate how many containers you need to buy to cover the 27 quarts that are required: 27 ÷ 5 = 5.4 quarts. It is not possible to buy a fractional part of a container, so you have to buy 6 containers.

76) The correct answer is D. Step 1 – Calculate the volume of each vat: length × width × height = 10 × 10 × 12 = 1,200 cubic feet. Step 2 – Determine how full each vat is in terms of cubic feet. Vat 1: 1,200 × $3/4$ = 1,200 × 0.75 = 900 cubic feet. Vat 2: 1,200 × $4/5$ = 1,200 × 0.80 = 960 cubic feet. Step 3 – Add the volume of the two vats together to determine the total volume: 900 + 960 = 1,860 cubic feet. Step 4 – Convert the cubic feet to cubic inches. 1 cubic foot = 1,728 cubic inches, so we multiply to convert: 1,860 cubic feet × 1,728 = 3,214,080 cubic inches. Step 5 – Multiply by the price to solve: 3,214,080 cubic inches × $0.12 = $385,689.60, which we round to $385,690.

77) The correct answer is B. Step 1 – Calculate the radius of the cone. The diameter is 6 and radius is half of diameter, so the radius is 3. Step 2 – Calculate the correct volume of the cone. The formula for the volume of a cone ≈ (3.14 × radius^2 × height) ÷ 3 = (3.14 × 3^2 × 8) ÷ 3 = 226.08 ÷ 3 = 75.36 cubic feet. Step 3 – Compare the correct figure to the erroneous figure to determine whether the erroneous calculation was too large or too small. You calculated 226 cubic feet, so you erred on the large side. Step 4 – Identify where the error occurred. We can see from the calculation in step 2 that final part of the calculation of the volume is (3.14 × 3^2 × 8) ÷ 3 = 226.08 ÷ 3, so you have forgotten to divide by 3.

78) The correct answer is B. No lights are to be installed in the corners, so each of the two 10-feet walls will have 1 light installed in the middle of each wall: 10 ÷ 5 = 2, but we subtract 1 from this for the corner. So, we have 1 light on each of the 2 shorter walls, which accounts for 2 lights so far. Each of the 25-foot walls have 5 increments of 5 feet, and again no lights are in the corners: (25 ÷ 5) – 1 = 4. So, each of the 2 long walls will have 4 lights on each wall. So there will be 10 lights in total on the walls in the room (1 + 1 + 4 + 4 = 10). You may wish to draw a diagram on your scratch paper when solving problems like this one.

79) The correct answer is D. Step 1 – Calculate the volume of the large ice cube: (1.8 × 1.8 × 1.8) = 5.832. Step 2 – Calculate the volume of the small ice cube: (1.4 × 1.4 × 1.4) = 2.744. Step 3 – Calculate the difference between the volume of the two ice cubes: 5.832 – 2.744 = 3.088

80) The correct answer is D. Step 1 – Calculate the area of the large triangle: (12 × 18) ÷ 2 = 216 ÷ 2 = 108. Step 2 – Calculate the area of the small triangle: (8 × 14) ÷ 2 = 112 ÷ 2 = 56. Step 3 – Subtract to solve: 108 – 56 = 52

81) The correct answer is D. To find the mean, add up all of the items in the set and then divide by the number of items in the set. Here we have 7 numbers in the set, so we get our answer as follows: (89 + 65 + 75 + 68 + 82 + 74 + 86) ÷ 7 = 539 ÷ 7 = 77

82) The correct answer is A. The mode is the number that occurs the most frequently in the set. Our data set is: 1, 1, 3, 2, 4, 3, 1, 2, 1. The number 1 occurs 4 times in the set, which is more frequently than any other number in the set, so the mode is 1.

83) The correct answer is B. The problem provides the number set: 8.19, 7.59, 8.25, 7.35, 9.10
First of all, put the numbers in ascending order: 7.35, 7.59, 8.19, 8.25, 9.10. Then find the one that is in the middle: 7.35, 7.59, **8.19**, 8.25, 9.10

84) The correct answer is A. Put the numbers is ascending order: 2, 2, 3, 5, **6, 8**, 8, 10, 12, 21. Here, we have got an even number of items, so we need to take an average of the two items in the middle:
(8 + 6) ÷ 2 = 7

85) The correct answer is C. To calculate the range, the low number in the set is deducted from the high number in the set. The problem set is: 98.5, 85.5, 80.0, 97, 93, 92.5, 93, 87, 88, 82. The high number is 98.5 and the low number is 80, so the range is 18.5 (98.5 − 80 = 18.5).

86) The correct answer is D. The mode is the number that occurs most frequently. However, if no number occurs more than once, the set has no mode.

87) The correct answer is B. We don't know the age of the 10th car, so put this in as x to solve:
(2 + 3 + 4 + 5 + 6 + 7 + 9 + 10 + 12 + x) ÷ 10 = 6
[(2 + 3 + 4 + 5 + 6 + 7 + 9 + 10 + 12 + x) ÷ 10] × 10 = 6 × 10
2 + 3 + 4 + 5 + 6 + 7 + 9 + 10 + 12 + x = 60
58 + x = 60
x = 2

88) The correct answer is A. Find the total points for the first group: 50 × 82 = 4100. Then find the total points for the second group. 50 × 89 = 4450. Add these two amounts together for the total points: 4100 + 4450 = 8550. Then divide the total points by the total number of members: 8550 ÷ 100 = 85.5

89) The correct answer is C. First, multiply the erroneous average by the erroneous number of tests to get the total points: 78 × 8 = 624. Then divide this total by the correct amount: 624 ÷ 10 = 62.4

90) The correct answer is A. Find the total of the items in the sample space: 5 + 10 + 8 + 12 = 35. We want to know the chance of getting an orange balloon, so put that in the denominator: $\frac{10}{35} = \frac{2}{7}$

91) The correct answer is D. We have 54 cards in the deck (13 × 4 = 52). We have taken out two spades, one heart, and a club, thereby removing 4 cards. So, the available data set is 48 (52 − 4 = 48). The desired outcome is drawing a heart. We have 13 hearts to begin with and one has been removed, so there are 12 hearts left. So, the probability of drawing a heart is $^{12}/_{48} = ^{1}/_{4}$

92) The correct answer is A. To solve problems like this one, it is usually best to write out the possible outcomes in a list. This will help you visualize the number of possible outcomes that make up the sample space. Then circle or highlight the events from the list to get your answer. In this case, we have two items, each of which has a variable outcome. There are 6 numbers on the black die and 6 numbers on the red die. Using multiplication, we can see that there are 36 possible combinations: 6 × 6 = 36
To check your answer, you can list the possibilities of the various combinations:

(1,1) (1,2) (1,3) (1,4) (1,5) (1,6)
(2,1) (2,2) (2,3) (2,4) (2,5) (2,6)
(3,1) (3,2) (3,3) (3,4) (3,5) (3,6)
(4,1) (4,2) (4,3) (4,4) (4,5) (4,6)
(5,1) (5,2) (5,3) **(5,4)** (5,5) (5,6)
(6,1) (6,2) (6,3) (6,4) (6,5) (6,6)

If the number on the left in each set of parentheses represents the black die and the number on the right represents the red die, we can see that there is one chance that Sam will roll a 4 on the red die and a 5 on the black die. The result is expressed as a fraction, with the event (chance of the desired outcome) in the numerator and the total sample space (total data set) in the denominator. So, the answer is 1/36.

93) The correct answer is D. You need to determine the number of possible outcomes at the start of the day first of all. The owner has 10 brown teddy bears, 8 white teddy bears, 4 black teddy bears, and 2 pink teddy bears when she opens the attraction at the start of the day. So, at the start of the day, she has 24 teddy bears: 10 + 8 + 4 + 2 = 24. Then you need to reduce this amount by the quantity of items that have been removed. The problem tells us that she has given out a brown teddy bear, so there are 23 teddy bears left in the sample space: 24 − 1 = 23. The event is the chance of the selection of a pink teddy bear. We know that there are two pink teddy bears left after the first prize winner receives his or her prize. Finally, we need to put the event (the number representing the chance of the desired outcome) in the numerator and the number of possible remaining combinations (the sample space) in the denominator. So the answer is 2/23.

94) The correct answer is C. Cobb County is the darkest bar, so it is the first bar for each month. For July, Cobb County had 1.2 inches of rain, and in August it had 0.8 inches, so it had 2 inches in total for the two months.

95) The correct answer is B. In June, Dawson County had 1.1 inches of rain and Emery County had 1.7 inches. Therefore, Emery County had 0.6 more inches of rainfall than Dawson County.

96) The correct answer is A. Emery County had the following amounts of rainfall: May = 2.5 inches; June = 1.8 inches; July = 1 inch; August = 0.9 inch. Then add these amounts together to solve: 2.5 + 1.8 + 1 + 0.9 = 6.2 inches in total

97) The correct answer is A. We already know that Emery County had 6.2 total inches of rainfall for the four months from our previous answer. So, add up the four months for Cobb County and then do the same for Dawson County. Cobb County: May = 2.9 inches; June = 2.1 inches; July = 1.2 inches; August 0.8 inches = 7 total inches. Dawson County: May = 3.5 inches; June = 1.1 inches; July = 0.9 inches; August = 2.3 inches = 7.8 total inches. Therefore, Emery County had the lowest total with 6.2 inches.

98) The correct answer is D. Reptiles account for 42% of the zoo creatures at the start of the year, and there are 1,500 creatures in total, so multiply to solve: 1,500 × 0.42 = 630 reptiles

99) The correct answer is B. At the start of the year, 26% of the zoo creatures were quadrupeds and 15% of the creatures were fish. So, solve by multiplying and subtracting as follows: (1500 × 0.26) − (1500 × 0.15) = 390 − 225 = 165 more quadrupeds than fish

100) The correct answer is C. We have to calculate the percentage of birds at the start of the year by subtracting the percentages for the other categories: 100% − 40% − 21% − 16% = 23%. The percentage of birds was 17% at the start of the year and 23% at the end of the year, so there was a 6% increase in the bird population. We can then multiply to solve: 1,500 × 0.06 = 90 more birds at the end of the year

101) The correct answer is C. Substitute −2 for x to solve.
$2x^2 - x + 5 =$
$[2 \times (-2^2)] - (-2) + 5 =$
$[2 \times (4)] - (-2) + 5 =$

(2 × 4) + 2 + 5 =
8 + 2 + 5 = 15

102) The correct answer is B. Isolate x to solve. You do this by doing the same operation on each side of the equation.
$-6x + 5 = -19$
Subtract 5 from each side to get rid of the integer 5 on the left side.
$-6x + 5 - 5 = -19 - 5$
Then simplify.
$-6x = -24$
Then divide each side by -6 to isolate x.
$-6x \div -6 = -24 \div -6$
$x = -24 \div -6$
$x = 4$

103) The correct answer is B.
Remember to do multiplication on the items in parentheses first.
$4x - 3(x + 2) = -3$
$4x - 3x - 6 = -3$
Then deal with the integers.
$4x - 3x - 6 + 6 = -3 + 6$
$4x - 3x = 3$
Then solve for x.
$4x - 3x = 3$
$x = 3$

104) The correct answer is C. Isolate the integers to one side of the equation.

$\frac{3}{4}x - 2 = 4$

$\frac{3}{4}x - 2 + 2 = 4 + 2$

$\frac{3}{4}x = 6$

Then get rid of the fraction by multiplying both sides by the denominator.

$\frac{3}{4}x \times 4 = 6 \times 4$

$3x = 24$

Then divide to solve the problem.

$3x \div 3 = 24 \div 3$

$x = 8$

105) The correct answer is B. Substitute 1 for x: $\frac{x-3}{2-x} = \frac{1-3}{2-1} = (1-3) \div (2-1) = -2 \div 1 = -2$

106) The correct answer is B.
Substitute 5 for the value of x to solve.
$x^2 + xy - y = 41$
$5^2 + 5y - y = 41$
$25 + 5y - y = 41$
$25 - 25 + 5y - y = 41 - 25$
$5y - y = 16$
$4y = 16$
$4y \div 4 = 16 \div 4$
$y = 4$

107) The correct answer is B. Substitute 12 for the value of x. Then simplify and solve.
$x^2 + xy - y = 254$
$12^2 + 12y - y = 254$
$144 + 12y - y = 254$
$144 - 144 + 12y - y = 254 - 144$
$12y - y = 110$
$11y = 110$
$11y \div 11 = 110 \div 11$
$y = 10$

108) The correct answer is C.
$6 + 8(2\sqrt{x} + 4) = 62$
$6 - 6 + 8(2\sqrt{x} + 4) = 62 - 6$
$8(2\sqrt{x} + 4) = 56$
$16\sqrt{x} + 32 = 56$
$16\sqrt{x} + 32 - 32 = 56 - 32$
$16\sqrt{x} = 24$
$16\sqrt{x} \div 16 = 24 \div 16$
$\sqrt{x} = 24 \div 16$
$\sqrt{x} = \frac{24}{16}$
$\sqrt{x} = \frac{24 \div 8}{16 \div 8} = \frac{3}{2}$

109) The correct answer is D. The factors of 50 are: 1 × 50 = 50; 2 × 25 = 50; 5 × 10 = 50. If any of your factors are perfect squares, you can simplify the radical. 25 is a perfect square, so, you need to factor inside the radical sign as shown to solve the problem: $\sqrt{50} = \sqrt{25 \times 2} = \sqrt{5^2 \times 2} = \sqrt{5^2} \times \sqrt{2} = 5\sqrt{2}$

110) The correct answer is D. 36 is the common factor, So, factor the amounts inside the radicals and simplify:

$\sqrt{36} + 4\sqrt{72} - 2\sqrt{144} =$

$\sqrt{36} + 4\sqrt{36 \times 2} - 2\sqrt{36 \times 4} =$

$\sqrt{6 \times 6} + 4\sqrt{(6 \times 6) \times 2} - 2\sqrt{(6 \times 6) \times 4} =$
$6 + (4 \times 6)\sqrt{2} - (2 \times 6)\sqrt{4} =$
$6 + 24\sqrt{2} - (12 \times 2) =$
$6 + 24\sqrt{2} - 24 =$
$-18 + 24\sqrt{2}$

111) The correct answer is A. $\sqrt{7} \times \sqrt{11} = \sqrt{7 \times 11} = \sqrt{77}$

112) The correct answer is B. Add the numbers in front of the radical signs to solve. If there is no number before the radical, then put in the number 1 because then the radical will count only 1 time when you add.
$\sqrt{15} + 3\sqrt{15} = 1\sqrt{15} + 3\sqrt{15} = (1 + 3)\sqrt{15} = 4\sqrt{15}$

113) The correct answer is B. The cube root is the number which satisfies the equation when multiplied by itself two times: $\sqrt[3]{\frac{216}{27}} = \sqrt[3]{\frac{6 \times 6 \times 6}{3 \times 3 \times 3}} = \frac{6}{3} = 2$

114) The correct answer is A. The base number is 7. Add the exponents: $7^5 \times 7^3 = 7^{(5+3)} = 7^8$

115) The correct answer is B. The base is xy. Subtract the exponents: $xy^6 \div xy^3 = xy^{(6-3)} = xy^3$

116) The correct answer is C. Perform the operation on the radicals and then simplify.
$\sqrt{8x^4} \cdot \sqrt{32x^6} = \sqrt{8x^4 \times 32x^6} = \sqrt{256x^{10}} = \sqrt{16 \times 16 \times x^5 \times x^5} = 16x^5$

117) The correct answer is B. We have the base number of 10 and we are multiplying, so we can add the exponent of 5 to the exponent of −1: (1.7 × 10^5 miles per hour) × (2 × 10^{-1} hours) = 1.7 × 2 × $10^{(5 + -1)}$ = 3.4 × 10^4 = 3.4 × 10,000 = 34,000 miles

118) The correct answer is D. When you have a fraction as an exponent, the numerator is new exponent and the denominator goes in front as the root: $\sqrt[7]{x^5} = (\sqrt[7]{x})^5$

119) The correct answer is B. The principle is that $x^{-b} = \frac{1}{x^b}$. Accordingly, $x^{-5} = \frac{1}{x^5}$

120) The correct answer is B. Following the principle mentioned in the answer to question 119, $(-4)^{-3} = \frac{1}{-4^3} = -\frac{1}{64}$

121) The correct answer is C. Any non-zero number raised to the power of zero is equal to 1.

122) The correct answer is C. Any non-zero number multiplied by a variable and raised to the power of zero is equal to 1.

123) The correct answer is C.

$$\frac{b + \frac{2}{7}}{\frac{1}{b}} = \left(b + \frac{2}{7}\right) \div \frac{1}{b} = \left(b + \frac{2}{7}\right) \times \frac{b}{1} = b\left(b + \frac{2}{7}\right) = b^2 + \frac{2b}{7}$$

124) The correct answer is D. Find the lowest common denominator for the second fraction. Then add the numerators.

$$\frac{x^2}{x^2 + 2x} + \frac{8}{x} = \frac{x^2}{x^2 + 2x} + \left(\frac{8}{x} \times \frac{x + 2}{x + 2}\right) = \frac{x^2}{x^2 + 2x} + \frac{8x + 16}{x^2 + 2x} = \frac{x^2 + 8x + 16}{x^2 + 2x}$$

125) The correct answer is A. Multiply as shown: $\frac{2a^3}{7} \times \frac{3}{a^2} = \frac{2a^3 \times 3}{7 \times a^2} = \frac{6a^3}{7a^2}$

Then find the greatest common factor and cancel out to simplify: $\frac{6a^3}{7a^2} = \frac{6a \times a^2}{7 \times a^2} = \frac{6a \times \cancel{a^2}}{7 \times \cancel{a^2}} = \frac{6a}{7}$

126) The correct answer is B. Invert and multiply.

$$\frac{8x + 8}{x^4} \div \frac{5x + 5}{x^2} = \frac{8x + 8}{x^4} \times \frac{x^2}{5x+5} = \frac{(8x \times x^2) + (8 \times x^2)}{(x^4 \times 5x) + (x^4 \times 5)} = \frac{8x^3 + 8x^2}{5x^5 + 5x^4}$$

Then factor out (x + 1) from the numerator and denominator and cancel out:

$$\frac{8x^3 + 8x^2}{5x^5 + 5x^4} = \frac{(8x^2 \times x) + (8x^2 \times 1)}{(5x^4 \times x) + (5x^4 \times 1)} = \frac{8x^2(x+1)}{5x^4(x+1)} = \frac{8x^2\cancel{(x+1)}}{5x^4\cancel{(x+1)}} = \frac{8x^2}{5x^4}$$

Finally, factor out x^2 and cancel it out: $\frac{8x^2}{5x^4} = \frac{x^2 \times 8}{x^2 \times 5x^2} = \frac{\cancel{x^2} \times 8}{\cancel{x^2} \times 5x^2} = \frac{8}{5x^2}$

127) The correct answer is A. The factors of 9 are: 1 × 9 = 9; 3 × 3 = 9. The factors of 3 are: 1 × 3 = 3. So, put the integer for the common factor outside the parentheses first: $9x^3 - 3x = 3(3x^3 - x)$
Then determine if there are any common variables for the terms that remain in the parentheses.

For $(3x^2 - x)$ the terms $3x^2$ and x have the variable x in common. So, now factor out x to solve:
$3(3x^3 - x) = 3x(3x^2 - 1)$

128) The correct answer is B. Looking at this expression, we can see that each term contains x. We can also see that each term contains y. So, first factor out xy: $2xy - 6x^2y + 4x^2y^2 = xy(2 - 6x + 4xy)$. We can also see that all of the terms inside the parentheses are divisible by 2. Now let's factor out the 2. To do this, we divide each term inside the parentheses by 2: $xy(2 - 6x + 4xy) = 2xy(1 - 3x + 2xy)$

129) The correct answer is C. The line in a fraction is the same as the division symbol. For example, $a/b = a \div b$. In the same way, $3/xy = 3 \div (xy)$.

130) The correct answer is A. You should use the FOIL method in this problem. Be very careful with the negative numbers when doing the multiplication.
$2(x + 2)(x - 3) =$
$2[(x \times x) + (x \times -3) + (2 \times x) + (2 \times -3)] =$
$2(x^2 + -3x + 2x + -6) =$
$2(x^2 - 3x + 2x - 6) =$
$2(x^2 - x - 6)$

Then multiply each term by the 2 at the front of the parentheses.
$2(x^2 - x - 6) =$
$2x^2 - 2x - 12$

131) The correct answer is A. To divide, invert the second fraction and then multiply as shown.
$\frac{x}{5} \div \frac{9}{y} = \frac{x}{5} \times \frac{y}{9} = \frac{x \times y}{5 \times 9} = \frac{xy}{45}$

132) The correct answer is D. Use the FOIL method to expand the polynomial.
FIRST – Multiply the first term from the first set of parentheses by the first term from the second set of parentheses: $(\mathbf{x} + 4y)(\mathbf{x} + 4y) = x \times x = x^2$
OUTSIDE – Multiply the first term from the first set of parentheses by the second term from the second set of parentheses: $(\mathbf{x} + 4y)(x + \mathbf{4y}) = x \times 4y = 4xy$
INSIDE – Multiply the second term from the first set of parentheses by the first term from the second set of parentheses: $(x + \mathbf{4y})(\mathbf{x} + 4y) = 4y \times x = 4xy$
LAST– Multiply the second term from the first set of parentheses by the second term from the second set of parentheses: $(x + \mathbf{4y})(x + \mathbf{4y}) = 4y \times 4y = 16y^2$
Finally, we add all of the products together: $x^2 + 4xy + 4xy + 16y^2 = x^2 + 8xy + 16y^2$

133) The correct answer is D. $(2 + \sqrt{6})^2 = (2 + \sqrt{6})(2 + \sqrt{6}) =$
$(2 \times 2) + (2 \times \sqrt{6}) + (2 \times \sqrt{6}) + (\sqrt{6} \times \sqrt{6}) = 4 + 4\sqrt{6} + 6 = 10 + 4\sqrt{6}$

134) The correct answer is C. As the studying increases, the grades also increase. A positive linear relationship therefore exists between the two variables. This is represented in chart C.

135) The correct answer is C. We can see that the line does not begin on exactly on (5, 5), nor does it begin on (5, 9) because the first point is slightly below the horizontal line for y = 5. Therefore, we can rule out answers A and D. If we look at x = 20 on the graph, we can see that y = 18 at this point. We can express this as the function: $f(x) = x \times 0.9$. Putting in the values of x from chart (C), we get the following: 5 × 0.9 = 4.5; 10 × 0.9 = 9; 15 × 0.9 =13.5; 20 × 0.9 = 18. This is represented in table C.

136) The correct answer is C. Put the value provided for x into the function to solve.
$f_1(2) = 5^x = 5^2 = 25$

137) The correct answer is D. First, solve for the function in the inner-most set of parentheses, in this case $f_1(x)$. To solve, you simply have to look at the first table. Find the value of 2 in the first column and the related value in the second column. For x = 2, $f_1(2)$ = 5. Then, take this new value to solve for $f_2(x)$.

Look at the second table. Find the value of 5 in the first column and the related value in the second column. For x = 5, $f_2(5)$ = 25.

138) The correct answer is C. Two whole numbers that are greater than 1 will always result in a greater number when they are multiplied by each other, rather than when those numbers are divided by each other or subtracted from each other. So, for positive integers, x × y will always be greater than the following:

x ÷ y

y ÷ x

x − y

y − x

1 ÷ x

1 ÷ y

139) The correct answer is B. Substitute $x + 3$ for x in the original function to solve. So, $x^2 + 3x - 8$ becomes $(x + 3)^2 + 3(x + 3) - 8$

140) The correct answer is B. First, you need to convert the logarithmic function into an exponential equation. To convert a logarithmic function to an exponent, the number after the equals sign (4 in this problem) becomes the exponent. The small subscript number after "log" (3 in this problem) becomes the base number. So, the exponential equation for $\log_3(x + 2) = 4$ is $3^4 = x + 2$. Then find the result for the exponent: 3^4 = 3 × 3 × 3 × 3 = 81. Substituting 81 on the left side of the equation, we get 81 = x + 2. Therefore, x = 79.

141) The correct answer is D. If a term or variable is subtracted within the parentheses, you have to keep the negative sign with it when you multiply.
FIRST: $(x - y)(x + y) = x \times x = x^2$
OUTSIDE: $(x - y)(x + y) = x \times y = xy$
INSIDE: $(x - y)(x + y) = -y \times x = -xy$
LAST: $(x - y)(x + y) = -y \times y = -y^2$
SOLUTION: $x^2 + xy + - xy - y^2 = x^2 - y^2$

142) The correct answer is A.
FIRST: $(3x + y)(x - 5y) = 3x \times x = 3x^2$
OUTSIDE: $(3x + y)(x - 5y) = 3x \times -5y = -15xy$
INSIDE: $(3x + y)(x - 5y) = y \times x = xy$
LAST: $(3x + y)(x - 5y) = y \times -5y = -5y^2$
Then add all of the above once you have completed FOIL: $3x^2 - 15xy + xy - 5y^2 = 3x^2 - 14xy - 5y^2$

143) The correct answer is A. First, Isolate the whole numbers.
$50 - \frac{3x}{5} \geq 41$
$(50 - 50) - \frac{3x}{5} \geq 41 - 50$
$-\frac{3x}{5} \geq -9$

Then get rid of the denominator on the fraction.

$-\frac{3x}{5} \geq -9$
$\left(5 \times -\frac{3x}{5}\right) \geq -9 \times 5$
$-3x \geq -9 \times 5$
$-3x \geq -45$
Then isolate the remaining whole numbers.
$-3x \geq -45$
$-3x \div 3 \geq -45 \div 3$
$-x \geq -45 \div 3$
$-x \geq -15$
Then deal with the negative number.
$-x \geq -15$
$-x + 15 \geq -15 + 15$
$-x + 15 \geq 0$
Finally, isolate the unknown variable as a positive number.
$-x + 15 \geq 0$
$-x + x + 15 \geq 0 + x$
$15 \geq x$
$x \leq 15$

144) The correct answer is D. Substitute 5 for $x - 2$ as shown: $x - 2 > 5$ and $x - 2 = y$, so $y > 5$. If two toys are being purchased, we need to solve for $2y$:

$y \times 2 > 5 \times 2$

$2y > 10$

145) The correct answer is B. For quadratic inequality problems like this one, you need to factor the inequality first. The factors of -9 are: -1×9; -3×3; 1×-9. Because we do not have a term with only the x variable, we need factors that add up to zero, so factor as shown:
$x^2 - 9 < 0$
$(x + 3)(x - 3) < 0$
Then find values for x by solving each parenthetical for 0.
$(x + 3) = 0$
$(-3 + 3) = 0$
$x = -3$

$(x - 3) = 0$
$(3 - 3) = 0$
$x = 3$
So, $x > -3$ or $x < 3$

You can then check your work to be sure that you have the inequality signs pointing the right way.
Use -2 to check $x > -3$. Since $-2 > -3$ is correct, our proof should also be correct:
$x^2 - 9 < 0$
$-2^2 - 9 < 0$
$4 - 9 < 0$
$-5 < 0$ CORRECT

Use 4 to check for $x < 3$. Since $4 < 3$ is incorrect, our proof should also be incorrect.
$x^2 - 9 < 0$
$4^2 - 9 < 0$
$16 - 9 < 0$
$7 < 0$ INCORRECT

Therefore, we have checked that $x > -3$ or $x < 3$.

146) The correct answer is B.
Factor: $x^2 - 5x + 6 \leq 0$
$(x - 2)(x - 3) \leq 0$

Then solve each parenthetical for zero:
$(x - 2) = 0$
$2 - 2 = 0$
$x = 2$
$(x - 3) = 0$
$3 - 3 = 0$
$x = 3$

So, $2 \leq x \leq 3$

Now check. Use 1 to check to $2 \leq x$, which is the same as $x \geq 2$. Since 1 is not actually greater than or equal to 2, our proof for this should be incorrect.

$x^2 - 5x + 6 \leq 0$
$1^2 - (5 \times 1) + 6 \leq 0$
$1 - 5 + 6 \leq 0$
$-4 + 6 \leq 0$
$2 \leq 0$ INCORRECT

Use 2.5 to check for $x \leq 3$. Since 2.5 really is less than 3, our proof should be correct.
$x^2 - 5x + 6 \leq 0$
$2.5^2 - (5 \times 2.5) + 6 \leq 0$
$6.25 - 12.5 + 6 \leq 0$
$-0.25 \leq 0$ CORRECT

Therefore, we have checked that $2 \leq x \leq 3$

147) The correct answer is D. We know that the products of 12 are: $1 \times 12 = 12$; $2 \times 6 = 12$; $3 \times 4 = 12$. So, add each of the two factors together to solve the first equation: $1 + 12 = 13$; $2 + 6 = 8$; $3 + 4 = 7$. (3, 4) solves both equations, so it is the correct answer.

148) The correct answer is C. The first term of the second equation is x. To eliminate the x variable, we need to multiply the second equation by 3 because the first equation contains $3x$.
$x + 2y = 8$
$(3 \times x) + (3 \times 2y) = (3 \times 8)$
$3x + 6y = 24$
Now subtract the new second equation from the original first equation.
$\quad 3x + 3y = 15$
$\underline{-(3x + 6y = 24)}$
$\quad\quad\quad -3y = -9$
Then solve for y.
$-3y = -9$
$-3y \div -3 = -9 \div -3$
$y = 3$
Using our original second equation of $x + 2y = 8$, substitute the value of 3 for y to solve for x.
$x + 2y = 8$
$x + (2 \times 3) = 8$
$x + 6 = 8$
$x + 6 - 6 = 8 - 6$
$x = 2$

149) The correct answer is B. There is a difference of 4 between each number in the sequence. Where variable a represents your starting number and variable d represents the difference, you could write an arithmetic sequence like this: a, a + d, a + 2d, a + 3d, a + 4d, a + 5d, . . .
However, it is easier to remember that the formula for the nth number of an arithmetic sequence is:

a + [d × (n – 1)] We can prove that 21 is the sixth number of the sequence in our problem by putting the values into the formula.
a = 1
d = 4
n = 6
a + [d × (n – 1)]
1 + [4 × (6 – 1)] =
1 + (4 × 5) =
1 + 20 = 21

150) The correct answer is D. Each number in the sequence is found by multiplying by a factor of 3:

2 × 3 = 6

6 × 3 = 18

18 × 3 = 54

So, each subsequent number is found by multiplying the previous number by 3. Where the first number is represented by variable a and the factor (called the "common ratio") is represented by variable r, you could write out a geometric sequence like this: $a, ar, a(r)^2, a(r)^3$ . . .

The sequence in this problem starts at 2 and triples each time, so a = 2 (the first term) and r = 3 (the "common ratio"). Remember that the formula for calculating the nth item in a geometric sequence is as follows: $ar^{(n-1)}$

So, let's consider our example problem again.

2, 6, 18, 54, . . .

The fifth term of the sequence is 54 × 3 = 162.

We can check this by putting the values into our formula: $ar^{(n-1)}$

a = 2 (the first term)

r = 3 (the "common ratio")

n = 5

$2 \times 3^{(5-1)}$ =

2×3^4 =

2 × 81

MATH FORMULA SHEET

Weight
1 ounce ≈ 28.350 grams
1 pound = 16 ounces
1 pound ≈ 453.592 grams
1 milligram = 0.001 grams
1 kilogram = 1,000 grams
1 kilogram ≈ 2.2 pounds

Volume
1 cup = 8 fluid ounces
1 quart = 4 cups
1 gallon = 4 quarts
1 gallon = 231 cubic inches
1 liter ≈ 0.264 gallons
1 cubic foot = 1,728 cubic inches
1 cubic yard = 27 cubic feet

Distance
1 foot = 12 inches
1 yard = 3 feet
1 mile = 5,280 feet
1 mile ≈ 1.61 kilometers
1 inch = 2.54 centimeters
1 foot = 0.3048 meters
1 meter = 1,000 millimeters
1 meter = 100 centimeters
1 kilometer = 1,000 meters
1 kilometer ≈ 0.62 miles

Area
1 square foot = 144 square inches
1 square yard = 9 square feet

Circle
number of degrees in circle = 360°
circumference ≈ 3.14 × *diameter*
area ≈ 3.14 × (*radius*)^2

Triangle
sum of angles = 180°
area = ½ (*base* × *height*)

Rectangle
perimeter = 2(*length* + *width*)
area = *length* × *width*

Rectangular Solid (Box)
volume = *length* × *width* × *height*

Cube
volume = (*length of side*)³

Cylinder
volume ≈ 3.14 × (*radius*)² × *height*

Cone
volume ≈ (3.14 × *radius*² × *height*) ÷ 3

Sphere (Ball)
volume ≈ 4/3 × 3.14 × *radius*

www.ingramcontent.com/pod-product-compliance
Lightning Source LLC
Chambersburg PA
CBHW081749100526
44592CB00015B/2351